Lecture Notes in Mathematics

Edited by A. Dold, Heidelberg and B. Eckmann, Zürich
Series: Tata Institute of Fundamental Research, Bombay
Adviser: M. S. Narasimhan

366

Robert Steinberg

University of California, Los Angeles, CA/USA

T0224623

Conjugacy Classes in Algebraic Groups

Notes by Vinay V. Deodhar

Springer-Verlag
Berlin · Heidelberg · New York 1974

AMS Subject Classifications (1970): 14 Lxx, 20-02, 20-G-xx

ISBN 3-540-06657-8 Springer-Verlag Berlin · Heidelberg · New York
ISBN 0-387-06657-8 Springer-Verlag New York · Heidelberg · Berlin

Offsetdruck: Julius Beltz, Hemsbach/Bergstr.

INTRODUCTION

The following is the substance of a set of lectures given at the Tata Institute of Fundamental Research during November and December of 1972. The notes are divided roughly into two parts. The first part attempts an a priori development of the basic properties of affine algebraic groups with emphasis on those needed in the study of conjugacy classes of elements of reductive groups : the semi-simple-unipotent decomposition, conjugacy of Borel subgroups and of maximal tori, completeness of the variety of Borel subgroups, etc. The second part is devoted to the classification and characterization of various such classes of elements: semisimple, unipotent, regular, subregular, etc. For a more detailed outline the reader may consult the table of contents. All of this is over an algebraically closed field. I had planned to include two talks on rationality questions, but this aim was not realized. Because of time limitations there had to be gaps in the actual development. In the first part the most serious of these is the omission of a large part of the proof of the existence of a quotient of a group by a closed subgroup. Also the principal structural and conjugacy results about connected solvable groups are used without proof, but this is not so serious since the Lie-Kolchin theorem <u>is</u> proved and from there on the proofs, by induction, follow fairly classical lines. In the second part the Bruhat lemma for reductive groups is used without proof (but a fairly complete proof is indicated for the classical groups) as are various properties of root systems and reflection groups (for which a comprehensive treatment may be found in Bourbaki's book). Modulo a few other points left to be checked by the reader I have attempted a coherent development.

It is a pleasure to thank my colleagues at the Tata Institute, young and old, for their hospitality and friendship to my wife and me during our visit and for their stimulating influence on my talks. It is a special pleasure to be able to thank here Shri Vinay V. Deodhar who in addition has written up these notes.

Robert Steinberg
University of California

TABLE OF CONTENTS

Chapter I

Affine algebraic varieties, affine algebraic groups and their orbits

Throughout this chapter, k will denote an algebraically closed field.

1.1. Affine algebraic varieties. Let k^n denote the cartesian product of n copies of k. Classically, a subset V of k^n is called an algebraic set if it is the set of zeros of a set of polynomials in $k[X_1, \ldots, X_n]$. k^n itself, the circle $x^2 + y^2 = 1$, a line in space, etc. are examples of such sets.

But this notion is unsatisfactory since it is not intrinsic. Hence we define an (abstract) affine algebraic variety in the following way:

It is a pair (V, A), where V is a set and A is a k-algebra of functions on V with values in k. This pair satisfies the following properties:

(1) A is finitely generated as k-algebra.

(2) A separates points of V i.e. given $x \neq y \in V$, there exists $f \in A$ such that $f(x) \neq f(y)$.

(3) Every k-algebra homomorphism $\phi : A \longrightarrow k$ is the evaluation at a point $x \in V$ i.e. $\phi(f) = f(x)\ \forall f \in A$.

Remarks. By (2), the point $x \in V$ is uniquely determined by the evaluation at that point (to be denoted as e_x). Thus, the points of V are in one-one correspondence with the k-algebra homomorphisms of A into k.

Examples of affine algebraic varieties:

(1) $(k^n, k[X_1, \ldots, X_n])$. (It is called the affine space of dimension n).

(2) $V \subsetneq k^n$, an algebraic set in earlier sense, $A = k[X_1, ..., X_n]/_V$.

(3) Let A be a finitely generated k-algebra without nilpotent elements. Then there exists an integer $n \geq 0$ and an exact sequence : $0 \to I \to k[X_1, ..., X_n] \to A \to 0$. Let $V = \{(a_1, \ldots, a_n) \in k^n / g(a_1, \ldots, a_n) = 0 \ \forall g \in I\}$. Then (V, A) is an affine algebraic variety. (This is a consequence of Hilbert's Nullstellensatz: see corollary to lemma 1 of 1.13). In fact, as we shall prove later, any affine algebraic variety is obtained in this way.

1.2. <u>Morphisms of affine algebraic varieties.</u> Let $(U, A), (V, B)$ be affine algebraic varieties. Then a morphism $f : (U, A) \longrightarrow (V, B)$ is a map $f: U \longrightarrow V$ such that the associated map f^* defined by the composition with f, takes B into A. f^* is called the comorphism associated to f.

<u>Remarks.</u> (1) For $u \in U$, the point $f(u) \in V$ is given by: $e_{f(u)} = e_u \circ f^*$. Thus f is completely determined by f^*.

(2) If $f: (U, A) \longrightarrow (V, B)$ and $g : (V, B) \longrightarrow (W, C)$ are morphisms of affine algebraic varieties then so is $g \circ f : (U, A) \longrightarrow (W, C)$ and $(g \circ f)^* = f^* \circ g^*$.

1.3. <u>Subvarieties of affine algebraic varieties.</u> Let (V, A) be an affine algebraic variety and $V' \subsetneq V$. If $(V', A/_{V'})$ is an affine algebraic variety in its own right, then it is called a subvariety of (V, A). It can be easily seen that $(V', A/_{V'})$ is a subvariety if and only if V' is the set of zeros of a set of elements in A. (The novice should check this.)

1.4. <u>Principal open affine subsets</u>. Let (V, A) be an affine algebraic variety and $f \in A$. Then $V_f = \{x \in V \mid f(x) = e_x(f) \neq 0\}$ is called a principal open subset of V. It can be seen that (V_f, A_f) is an affine algebraic variety. Here $A_f = A\left[\frac{1}{f}\right]$.

1.5. A basic lemma. Here, we prove an important lemma which will be used quite often later on.

Lemma. Let (U,A), (V,B) be affine algebraic varieties and $f : (U, A) \longrightarrow (V, B)$ be a morphism. Let $f^*: B \longrightarrow A$ be the associated comorphism. If f^* is onto, then f(U) is an algebraic subvariety of V and $f : U \longrightarrow f(U)$ is an isomorphism.

Proof. Let $0 \longrightarrow I \longrightarrow B \xrightarrow{f^*} A \longrightarrow 0$ be exact.

Claim: $f(U) = \left\{ v \in V \,/\, e_v(g) = 0 \ \forall g \in I \right\}$. Let $u \in U$. Then for $g \in I$, $e_{f(u)}(g) = (e_u \circ f^*)(g) = 0$. Conversely, let $v \in V$ such that $e_v(g) = 0 \ \forall g \in I$. Clearly, e_v quotients to $\bar{e}_v : A \longrightarrow k$ such that $\bar{e}_v = e_v \circ f^*$. Now (U,A) is an affine algebraic variety, hence there exists $u \in U$ such that $\bar{e}_v = e_v \circ f^* = e_u$. Hence $v = f(u)$ as B separates V. This proves the claim. Hence f(U) is an algebraic subvariety of V and $B/_{f(U)} \xrightarrow{\sim} \frac{B}{I}$. Hence $(f(U), \frac{B}{I})$ is a subvariety of (V, B). Clearly, there exists $g^* : A \longrightarrow \frac{B}{I}$ such that $g^* \circ f^*$ and $f^* \circ g^*$ are respective identities. The morphism g defined by g^* is such that $g \circ f$ and $f \circ g$ are respective identities. Hence the lemma.

Proposition 1. Every abstract affine algebraic variety is isomorphic to a sub-variety of the affine algebraic variety $(k^n, k[X_1, ..., X_n])$ for suitable n.

Proof. Let (V, A) be an affine algebraic variety. A is finitely generated say by $f_1, \ldots, f_n$. Define $\phi : V \longrightarrow k^n$, given by: $\phi(v) = (f_1(v), \ldots, f_n(v))$. It can be easily seen that the corresponding map ϕ^* is given by $\phi^*(X_i) = f_i$ $1 \leq i \leq n$. Clearly ϕ^* maps $k[X_1, ..., X_n]$ onto A. Hence by the above lemma, $\phi(V)$ is a subvariety of k^n and $\phi : V \longrightarrow \phi(V)$ is an isomorphism. Hence the proposition.

1.6. Products of affine algebraic varieties. Let (U, A), (V, B) be affine algebraic varieties. Then elements of $A \otimes B$ can be treated as functions on $U \times V$. Explicitely, $(a \otimes b)(x, y) = a(x).\ b(y)$, $x \in U$, $y \in V$. Then $(U \times V, A \otimes B)$ can be seen to be an affine algebraic variety, called the product of (U, A) and (V, B). Here, only property (3) is to be varified. Let $\phi : A \otimes B \longrightarrow k$ be a k-algebra homomorphism. This gives rise to $\phi_1 : A \longrightarrow k$, $\phi_2 : B \longrightarrow k$ defined by : $\phi_1(a) = \phi(a \otimes 1)$ and $\phi_2(b) = \phi(1 \otimes b)$; $a \in A$, $b \in B$. Hence there exists $x \in U, y \in V$ such that $\phi_1 = e_x$, $\phi_2 = e_y$.

Now, $(a \otimes b)(x, y) = a(x).\ b(y) = e_x(a).\ e_y(b)$

$$= \phi_1(a)\ .\ \phi_2(b) = \phi(a \otimes b).$$

Thus $\phi = e_{(x,y)}$.

Again the maps $A \hookrightarrow A \otimes B$, $B \hookrightarrow A \otimes B$ give rise to morphisms $U \times V \xrightarrow{\pi_1} U$; $U \times V \xrightarrow{\pi_2} V$ which, in fact, are the projections.

$(U \times V, A \otimes B)$ has the following universal property: Given an affine algebraic variety (W, C) and morphisms $p_1 : (W, C) \longrightarrow (U, A)$ and $p_2 : (W, C) \longrightarrow (V, B)$, there exists a unique morphism $p : (W, C) \longrightarrow (U \times V, A \otimes B)$ such that $\pi_1 \circ p = p_1$, $\pi_2 \circ p = p_2$. This property follows immediately from a corresponding universal property in tensor products of commutative algebras or else can be verified directly. As an exercise the novice may wish to prove the important fact that each of the morphisms p_1, p_2 above is open (maps open sets onto open sets , in the Zariski topology defined in 1.10).

1.7. Notion of affine algebraic groups. An affine algebraic group is a pair (G, A) such that

(1) (G, A) is an affine algebraic variety

(2) G is a group

(3) The group operations are morphisms

i.e. $m : G \times G \longrightarrow G$, $m(x, y) = x.y$ and $i : G \longrightarrow G$, $i(x) = x^{-1}$

are morphisms.

Examples of affine algebraic groups:

(1) Let V be an n-dimensional vector space over k. Then

$(GL(V), k[T_{11}, \ldots, T_{nn}]_D)$ where D is the determinant of (T_{ij}) and

$k[T_{11}, \ldots, T_{nn}]_D$ is the ring obtained from $k[T_{11}, \ldots, T_{nn}]$ by adjoining

D^{-1} (This will be discussed in 1.9).

(2) SL(V) is an algebraic subvariety of GL(V) and is an affine algebraic group in its own right.

(3) The group of diagonal matrices in GL(V) as subvariety of GL(V) is an affine algebraic group.

(4) The invertible elements of any finite dimensional associative k-algebra.

(The groups in (1), (2) and (3) are called linear algebraic groups.)

A linear algebraic group is an affine subvariety of GL(V), for some finite dimensional vector space V, which is a subgroup also.

1.8. Comorphisms in affine algebraic groups. Let (G, A) be an affine algebraic group. Let $m : G \times G \longrightarrow G$ and $i : G \longrightarrow G$ be the group operations. These give rise to comorphisms $m^* : A \longrightarrow A \otimes A$ and $i^* : A \longrightarrow A$. Consider the morphism : $\rho_x : G \longrightarrow G$ given by $\rho_x(y) = y.x$ (= m(y, x)) for a fixed $x \in G$. This gives rise to a comorphism $\rho_y^*: A \longrightarrow A$. This in fact is a k-algebra automorphism of A, since ρ_x is an automorphism of G as a variety. Also, $\rho_{xy} = \rho_y \circ \rho_x$, hence $\rho_{xy}^* = \rho_x^* \circ \rho_y^*$. Thus, $\rho^*: G \longrightarrow$ Automorphism of A, $\rho^*(x) = \rho_x^*$ is a group homomorphism.

ρ_x^* is called the right translation of A by $x \in G$. Similarly λ_x^*, the left translation of A by x can be defined.

<u>Lemma</u>. Let B be the subspace of A spanned by $\{\rho_x^*(f),\ x \in G\}$ (for a fixed $f \in A$). Then B is finite dimensional.

<u>Proof</u>. If $f = 0$, there is nothing to prove. So let $f \neq 0$.

Let $m^*f = \sum_{i=1}^{n} g_i \otimes h_i,\ g_i, h_i \in A$ such that n is minimal.

<u>Claim</u>. $\{g_i\}_{1 \leq i \leq n}$ is a basis of B.

Firstly, g_i' s are linearly independent (from minimality of n). For the same reason, h_i's are also linearly independent. Hence there exist $\{x_j\}_{1 \leq j \leq n}$ in G such that $(h_i(x_j))_{(i,j)}$ has a non-zero determinant. Again, for any $y \in G$,

$$(\rho_y^* f)(x) = f(x.y) = \sum_{i=1}^{n} g_i(x) \cdot h_i(y).$$

Hence $\rho_y^* f = \sum_{i=1}^{n} h_i(y).\, g_i$ ———— I.

In particular, $\rho_{x_j}^* f = \sum_{i=1}^{n} h_i(x_j).g_i \quad 1 \leq j \leq n$.

Hence for each i, g_i is a linear combination of $(\rho_{x_j}^* f)_{1 \leq j \leq n}$.

Thus, $$g_i \in B \ \forall\ 1 \leq i \leq n.$$

From I above, B is spanned by $\{g_i\}_{1 \leq i \leq n}$ Hence the claim and the lemma.

<u>Remarks</u>. In the above lemma, one can take the left-translations $\lambda_G^* = \{\lambda_x^*, x \in G\}$ or the left and right translations $\rho_G^* \circ \lambda_G^*$ instead of the

right-translations ρ_G^*.

1.9. Linear algebraic groups. Let V be an n-dimensional vector-space/$_k$. For $v \in V$ and $a^* \in V^*$, consider the matrix coefficient $m_{v,a^*} : GL(V) \longrightarrow k$ defined by :

$$m_{v,a^*}(T) = a^*(T(v)).$$

Clearly, $$\left.\begin{aligned} m_{v+v',a^*} &= m_{v,a^*} + m_{v',a^*} \\ m_{v,a^*+b^*} &= m_{v,a^*} + m_{v,b^*} \end{aligned}\right\} \quad \begin{aligned} &v, v' \in V \\ &a^*, b^* \in V^* \end{aligned}$$

Let A' be the k-algebra generated by $\{m_{v,a^*} \mid v \in V, a^* \in V^*\}$. Let $D \in A'$ be the function : $GL(V) \longrightarrow k$ given by $D(T)$ = determinant of T (which is well-defined). Let A be the ring obtained from A' by adjoining D^{-1}. Then $(GL(V), A)$ is an affine algebraic group. Any subvariety of $(GL(V), A)$ which is also a subgroup is called a linear algebraic group.

Lemma. Let (G, A) be an affine algebraic group, $B \subsetneq A$ be a finite dimensional subspace invariant under ρ_G^*. Define $\alpha : G \longrightarrow GL(B)$ by $\alpha(x) = \rho_x^*/_B$. Then α is a morphism of affine algebraic groups and the matrix coefficients on G (pulled back via α) generate the same linear space as $\{\lambda_G^* B\}$.

Proof. Let $v \in D$, $a^* \in B^*$. Then $m_{v,a^*} : GL(B) \longrightarrow k$ is given by:

$$m_{v,a^*}(T) = a^*(T(v)).$$

Let $m^* v = \sum_{i=1}^{n} g_i \otimes h_i$ (n minimal). Then as proved earlier, $\{g_i\}_{1 \leq i \leq n}$ is a basis of the linear space spanned by the right-translates of v and $\{h_i\}_{1 \leq i \leq n}$ is a basis of the linear space spanned by the left-translates of v.

Now $(m_{v,a^*} \circ \alpha)(y) = a^*(\alpha(y)\,(f)) = a^*(\rho^*_y f)$

$$= a^*\left(\sum_1^n h_i(y).g_i\right) = \sum_1^n a^*(g_i).h_i(y)$$

or $$m_{v,a^*} \circ \alpha = \sum_{i=1}^n a^*(g_i).h_i \qquad \text{———} \quad \text{I.}$$

Again, $\dfrac{1}{(D \circ \alpha)(y)} = \dfrac{1}{\det.\ (\alpha(y))} = \det.\ \alpha(y^{-1}) = \det.\ (\alpha \circ i)(y)$. Hence

$\frac{1}{D} \circ \alpha = D \circ \alpha \circ i$. and $i : G \longrightarrow G$ is a morphism. Hence from I above, $\frac{1}{D} \circ \alpha \in A$.

<u>Thus:</u> (1) α is a morphism.

Next, g_i's are linearly independent. Hence choosing a^* properly, h_i can be shown to be in the subspace W spanned by the matrix coefficients. Thus the space generated by left-translates of $v \subseteq W$. From I, clearly, $W \subseteq$ linear space spanned by $\{\lambda^*_G B\}$. Hence the lemma.

We now prove a fundamental proposition:

<u>Proposition.</u> Any (abstract) affine algebraic group is isomorphic to a linear algebraic group.

<u>Proof.</u> Let (G, A) be an affine algebraic group. Then A is finitely generated, say by $f_1 = 1, \ldots, f_k \in A$. Let B be the linear space spanned by:

$$\left\{ \rho^*_x f_i \,\middle|\, x \in G,\ 1 \le i \le n \right\}.$$

Then by an earlier lemma, B is finite dimensional and obviously invariant under ρ^*_G. Hence the previous lemma applied to B gives a morphism $\alpha : G \longrightarrow GL(B)$ such that the linear space W spanned by $\lambda^*_G B$ is the same as the one spanned by matrix coefficients of G. But then the image of α^* is a k-subalgebra which contains W which, in turn, contains $\{f_i\}_{1 \le i \le n}$. Hence $A = \text{Image } \alpha^*$. The proposition now follows from the lemma in 1.5.

A final remark. Just as in the case of abstract algebraic varieties, so in the case of algebraic groups (G,A) everything can be retracted into A. The points of G are the k-algebra homomorphisms $x : G \longrightarrow k$. The multiplication in G is reflected by the comultiplication $m^* : A \longrightarrow A \otimes A$ ($f \rightsquigarrow \sum g_i \otimes h_i$ earlier); x and y are multiplied by: $(xy)(f) = \sum g_i(x) \cdot h_i(y)$. A thus becomes a coalgebra as well as an algebra with compatible comultiplication and multiplication (Hopf algebra) and other conditions to reflect the group laws.

1.10. Zariski-topology on varieties. Let (V,A) be an affine algebraic variety. Let $\mathcal{F} = \{F \subseteq V / F$ is the set of zeros of a set of functions in $A\}$ i.e. Elements of $\mathcal{F}$ are just the algebraic subsets of V. Now the following statements can be easily verified:

(1) $\emptyset, V \in \mathcal{F}$

(2) Let $F_\lambda \in \mathcal{F}\ \forall \lambda \in \Lambda$. Let F_λ be the set of zeros of $\{f_{\lambda\mu}\}_{\mu \in M_\lambda} \subseteq A$. Then $\cap F_\lambda$ is the set of zeros of $\{f_{\lambda\mu}\}_{\mu \in M_\lambda, \lambda \in \Lambda}$.

(3) Let $F_1, F_2 \in \mathcal{F}$. Let F_i be the set of zeros of $\{f^i_\lambda\}_{\lambda \in \Lambda_i}$; i = 1, 2. Then $F_1 \cup F_2$ is the set of zeros of $\{f^1_{\lambda_1} \cdot f^2_{\lambda_2}\}_{\lambda_1 \in \Lambda_1, \lambda_2 \in \Lambda_2}$.

Hence $\mathcal{F}$ defines a topology on V. (The closed sets are just the elements of $\mathcal{F}$). This topology is called the Zariski-topology on V. It can be verified that a morphism of varieties is continuous with respect to this topology.

We state here an important result, which follows from the 'Hilbert basis theorem'. The result is: A finitely generated k-algebra is noetherian, i.e. satisfies the maximal condition on ideals.

1.11. We now define the notions of noetherian space and irreducible space.

Let $(X, \mathfrak{T})$ be a topological space. We have the following definitions:

(1) $(X, \mathfrak{T})$ is said to be noetherian if it satisfies the minimal condition on closed sets i.e. every non-empty family of closed sets contains a minimal element. (It follows immediately that a space is noetherian iff any decreasing sequence of closed sets terminates iff any increasing sequence of open sets terminates iff a non-empty family of open sets has a maximal element).

(2) $(X, \mathfrak{T})$ is called irreducible if it is not a union of two proper closed subsets.

(3) A subset of a space is irreducible if it is irreducible in the subspace topology. (It can be easily seen that a subset A of a space X is irreducible iff for every U, V open in X, such that $U \cap A \neq \emptyset$, $V \cap A \neq \emptyset$, we have $U \cap V \cap A \neq \emptyset$. Thus in an irreducible space, every non-empty open set is dense).

Lemma. Let (V, A) be an affine algebraic variety. Then (V, A) is irreducible iff A is an integral domain.

Proof. Let (V, A) be irreducible and $f.g = 0$, for $f, g \in A$. Consider $S = \{x \in V / f(x) = 0\}$ and $T = \{x \in V/g(x) = 0\}$. Then S and T are closed in V and $V = S \cup T$. Since V is irreducible, $V = S$ or $V = T$. Hence $f = 0$ or $g = 0$. Conversly, let A be an integral domain. Consider the principal open sets V_f and V_g; $f \neq 0$, $g \neq 0$. Since $f.g \neq 0$, it follows that $V_f \cap V_g \neq \emptyset$. Further, any open set S contains a principal open set V_f, $f \in A$ and $f \neq 0$. Now it follows immediately that every two non-empty open sets intersect. Hence V is irreducible.

Remarks. (a) A morphic image of an irreducible variety is irreducible.

(b) If (U, A) and (V, B) are irreducible algebraic varieties, then so is $(U \times V, A \otimes B)$.

(Exercise: Prove this using (1), (2), (3) in the definition of a variety).

We now prove a basic theorem for noetherian spaces.

Theorem. Every noetherian space can be expressed as a finite union of closed irreducible subsets. If such an expression is irredundant, i.e. if no subset may be omitted, then the irreducible subsets involved are uniquely determined as the maximal irreducible subsets of V. (Such subsets are called irreducible components of V).

Proof. Let V be a noetherian space. Let $\mathcal{F} = \{F \subseteq V$ which are closed and such that F cannot be expressed as a union of finitely many closed irreducible subsets of $F\}$. If possible, let $\mathcal{F} \neq \emptyset$. So let $F \in \mathcal{F}$ which is minimal (V is noetherian). Now F itself cannot be irreducible. Hence $F = F_1 \cup F_2$; F_1, F_2 are proper closed subsets of V. The minimality of F in $\mathcal{F}$ shows that both F_1 and F_2 are expressible as finite unions of closed irreducible subsets; but then so is F, which is a contradiction since $F \in \mathcal{F}$. Hence $\mathcal{F} = \emptyset$ and V, in particular, is expressed as a finite union of closed irreducible subsets.

Consider an irredundant expression $V = F_1 \cup \ldots \cup F_r$. F_i is a closed irreducible subset $1 \leq i \leq r$. If possible, let $F_1 \subsetneq F$, F is a closed irreducible subset.

Then $F = F \cap V = (F \cap F_1) \cup \ldots \cup (F \cap F_r) = F_1 \cup (F \cap F_2) \cup \ldots \cup (F \cap F_r)$.

Now $F \cap F_i$ is closed $(i \geq 2)$ and F irreducible. Hence $F_1 \subseteq F \subseteq F_i$ for some $i \geq 2$. This contradicts the irredundancy of $V = F_1 \cup \ldots \cup F_r$. Hence F_1 is a maximal closed irreducible subset of V. Similarly $F_i (i \geq 2)$ also has the above property. Again, if F is a maximal closed irreducible subset, then $F = (F \cap F_1) \cup \ldots \cup (F \cap F_r)$ gives $F \subseteq F_i$ for some i, which in turn implies $F = F_i$. Hence all the maximal closed irreducible subsets of V occur exactly once in $F_1 \cup \ldots \cup F_r$. Hence the theorem.

Corollary. Let (V, A) be an affine algebraic variety. Then the above theorem is true for V, endowed with the Zariski-topology. (In fact, we prove: (V, Zariski) is noetherian).

Proof. As a consequence of Hilbert's basis theorem, we have: Every finitely generated k-algebra is noetherian. Hence A is noetherian i.e. every ideal of A is finitely generated (which is equivalent to saying: A satisfies the maximum condition on ideals).

With every closed subset U of V, we associate an ideal I(U) of A defined by : $I(U) = \{ f \in A / f(x) = 0 \ \forall x \in U \}$ i.e. I(U) is the ideal of all elements of A vanishing on U. Since U is closed, it follows that U is precisely the set of zeros of I(U). Thus, $U \underset{\neq}{\supset} U'$, U and U' closed, imply that $I(U) \underset{\neq}{\subset} I(U')$. Hence the maximum condition on ideals of A implies the minimum condition on the closed sets of V. Thus (V, Zariski) is noetherian. Hence the previous theorem is applicable.

1.12. Irreducible components of an affine algebraic group. Let (G, A) be an affine algebraic group. Since G has a group structure on it and the group operations are morphisms, the irreducible components of G have a special nature which is described in the following proposition:

Proposition 1. Let (G, A) be an affine algebraic group. Then the irreducible components of G are disjoint. If G^o is the irreducible component of G containing e, then G^o is a (closed) normal subgroup of G having a finite index in G. Further, the irreducible components of G are precisely the cosets of G^o.

Proof. Let, if possible, two components intersect. Hence $\exists\, x \in G$ such that x belongs to two distinct irreducible components. Since an automorphism of G as a variety permutes the irreducible components and since the left-translation by an element of G is an automorphism, it follows that every element of G belongs to two distinct irreducible components. Take an irreducible component V. Each of its elements belongs to an irreducible component other than V. This clearly contradicts the irredundancy of the expression of G as a union of its irreducible components. Hence the irreducible components of G are disjoint. Let $x \in G^o$; then xG^o is also a component and it contains x. Hence by disjointness, $xG^o = G^o$. Hence $G^o . G^o \subseteq G^o$. For a similar argument, ${G^o}^{-1} = G^o$, and for $y \in G$ arbitrary, $yG^oy^{-1} = G^o$. Hence G^o is a normal subgroup of G. Clearly, its cosets are also irreducible components of G. Conversly, let F be an irreducible component of G. Choose $x \in F$, then $x^{-1}F = G^o$ or $F = xG^o$. Hence the result.

Remark 1. G^o is the smallest closed subgroup of G having finite index in G. (Any closed subgroup of finite index is open also).

Remark 2. If S is a closed subsemigroup, then it is, in fact, a subgroup (exercise).

Remark 3. For an algebraic group, the irreducible components and connected components are the same. We call G^o the (connected) component of the identity e and call G connected if $G = G^o$.

Remark 4. As an example, we consider $G = O_n$. Here $G^o = SO_n$ so that G consists of two components. The groups GL_n, SL_n, Sp_{2n}, Diag, Superdiag, ... on the other hand are all connected.

1.13. Hilbert's second theorem. Until now k could have been any field, even a finite one. The assumption that k is algebraically closed will now be brought into play.

Notation. Henceforth, we denote an affine algebraic variety (V, A) by V. The algebra A of functions on V is not mentioned unless required, and is sometimes also written $k[V]$. Similarly, by a variety V we mean an affine algebraic variety.

We start with a definition. Let V be a variety and $U \subseteq V$. Then U is épais in V if

(1) U is irreducible

(2) U contains a dense open subset of $\overline{U}$.

The main proposition is :

Proposition 1. Let U, V be varieties ; $\alpha : U \longrightarrow V$ be a morphism. Let $U' \subseteq U$ be an épais. Then $\alpha(U')$ is an épais in V.

Proof. $\overline{U'}$ is a variety in its own right and $\alpha/\overline{U'}$ is a morphism. Hence, without loss of generality, one may assume that U' is dense in U. Further, U' contains an open dense set which is principal. (Since $\overline{U'} = U$ is irreducible, any open set is empty or dense). Hence, without loss of generality, U' itself may be assumed to be a principal open dense set. Let $U' = \{x \in U / f(x) \neq 0\}$ for some $f \in k[U]$. Again, $\overline{\alpha(U)}$ is a variety in its own right, and hence one may assume that $\alpha(U)$ is in fact dense in V. It follows that $\alpha^*: k[V] \longrightarrow k[U]$ is injective. Also, both of $k[V]$ and $k[U]$ are integral domains. (U and V are both irreducible). We now state a lemma which will be proved later.

Lemma 1. Let A and B be integral domains, $A \supseteq B$, and A finitely generated over B. Let $f \neq 0$, $f \in A$. Then there exists $g \neq 0$, $g \in B$ such that for any algebraically closed field F and a homomorphism $\alpha: B \longrightarrow F$ with $\alpha(g) \neq 0$, α extends to a homomorphism $\tilde{\alpha}: A \longrightarrow F$ such that $\tilde{\alpha}(f) \neq 0$.

We prove the proposition using this lemma. Take $A = k[U]$, $B = \alpha^*(k[V])$. Since $U' = \{x \in U/f(x) \neq 0\}$ is dense in U, $f \neq 0$. Hence by the lemma, $\exists$ $\alpha^*(g) \neq 0$, $\alpha^*(g) \in B$ $(g \in k[V])$ having the said properties. Let $W = \{y \in V/g(y) \neq 0\}$. Then W is open in V, $W \neq \emptyset$, hence dense in V.
Claim. $W \subseteq \alpha(U')$, which clearly proves the proposition. So let $q \in W$. Then $g(q) \neq 0$. Consider $e_q \circ \alpha^{*-1}: B \longrightarrow k$ which is well-defined since α^* is injective onto B. Also, $e_q \circ \alpha^{*-1}(\alpha^*(g)) = e_q(g) = g(q) \neq 0$. Hence by the lemma, $\exists\, \emptyset : A \longrightarrow k$ such that $\emptyset/_B = e_q \circ \alpha^{*-1}$ and $\emptyset(f) \neq 0$. But then $\emptyset = e_p$ for some $p \in U$ by property (3) of varieties and $\emptyset(f) = f(p) \neq 0$. Hence $p \in U'$. Also, for $h \in k[V]$, $h(\alpha(p)) = e_{\alpha(p)}(h) = e_p \circ \alpha^*(h) = \emptyset(\alpha^*(h)) = e_q \circ \alpha^{*-1}(\alpha^*(h)) = e_q(h) = h(q)$.
Hence $q = \alpha(p)$.

Hence $W \subseteq \alpha(U')$.

Proof of Lemma 1. The lemma is proved in several steps:

Step 1. Let F be a field, R be a subring of F and $x \neq 0 \in F$. Then every homomorphism of R into an algebraically closed field K can be extended to $R[x]$ or $R[x^{-1}]$.

Proof. Let $\alpha: R \longrightarrow K$ be a homomorphism. Let P = kernal of α. Then clearly P is a prime ideal. Consider R_P, the localisation of R at P. α gives rise to $\tilde{\alpha}: R_P \longrightarrow K$ defined by : $\tilde{\alpha}(a/s) = \dfrac{\alpha(a)}{\alpha(s)}$, $a \in R$, $s \in R - P$. If $\tilde{\alpha}$ can be extended to $R_P[x]$ or $R_P[x^{-1}]$, then clearly α is extended to $R[x]$ or $R[x^{-1}]$. Also, $\tilde{\alpha}(R_P)$ is a field. Hence, without loss of generality, we may assume that $\alpha(R)$ itself if a field, say $\bar{R}$. For a polynomial $g(X)$ with coefficients in R, let $\bar{g}(X)$ denote the polynomial $\in \bar{R}[X]$ obtained via α. Let $I = \{\bar{g}(X) \in \bar{R}[X] / g(x) = 0\}$. Clearly, I is an ideal in $\bar{R}[X]$ which is a P.I.D. Hence I is generated by some $\bar{g}_0(X)$.

(i) If $\bar{g}_0(X)$ is not a non-zero constant, then choose $\lambda \in K$ such that $\bar{g}_0(\lambda) = 0$ (K is algebraically closed). Define $\tilde{\alpha}: R[x] \longrightarrow K$ by $\tilde{\alpha} = \alpha$ on R and $\tilde{\alpha}(x) = \lambda$. Clearly, $\tilde{\alpha}$ extends uniquely to the whole of $R[x]$. Hence the step. ——— (I)

(ii) If $\bar{g}_0(X)$ is a non-zero constant, it can be assumed to be 1. Again, every element b of R such that $\alpha(b) \neq 0$ is invertible. Hence there exists $g_0(X) \in R[X]$ such that $g_0(x) = 0$ and g_0 is of the form:

$$g_0(X) = 1 + a_1 X + \ldots + a_m X^m, \; a_1, \ldots, a_m \in P. \qquad \text{——— II}$$

Now consider x^{-1} instead of x. Either I holds in which case α is extended to $R[x^{-1}]$ or there exists $f_0(Y) \in R[Y]$ such that

$$f_0(Y) = 1 + b_1 Y + \ldots + b_n Y^n, \; b_1, \ldots, b_n \in P \text{ and } f_0(\tfrac{1}{x}) = 0. \qquad \text{——— III}$$

Again, both g_o and f_o can be assumed to be of minimal degree amongst the polynomials satisfying above mentioned conditions (II and III). Assume, without loss of generality, $n \leq m$.

Now $$0 = 1 + \frac{b_1}{x} + .. + \frac{b_n}{x^n} .$$

Thus $$0 = x^m + b_1 x^{m-1} + ... + b_n . x^{m-n}$$

and $$0 = 1 + a_1 x + ... + a_{m-1} x^{m-1} + a_m(-b_1 x^{m-1} - .. - b_n x^{m-n})$$

and this contradicts the minimality of m. This contradiction proves the step.

Note that if $x \in R$ and $\alpha(x) \neq 0$, then from III, it follows that α extends to $R[x^{-1}]$. For, if not, (III) holds.

Hence $$f_o(Y) = 1 + b_1 Y + .. + b_n Y^n, b_1, ..., b_n \in P, f_o(\tfrac{1}{x}) = 0.$$

But then $0 = x^n + b_1 x^{n-1} + ... + b_n$. Hence $\alpha(x)^n = 0$ which contradicts $\alpha(x) \neq 0$.

Step 2. Let F, R be as in Step 1. Let R_o be a maximal subring of F such that $R_o \supseteq R$ and α extends to R_o. Then R_o is a valuation ring, i.e. $x \in F, x \neq 0 \Rightarrow x \in R_o$ or $x^{-1} \in R_o$.

Proof. Let $\mathfrak{J} = \{(R', \alpha')/R'$ subring of $F, R' \supseteq R$ and α' is an extension of $\alpha\}$. On $\mathfrak{J}$, define an order by : $(R', \alpha') \geq (R'', \alpha'')$ if $R' \supseteq R''$ and $\alpha'' = \alpha'/R''$. Then every chain in $\mathfrak{J}$ is bounded above. Hence by Zorn's lemma, $\mathfrak{J}$ has a maximal element, say (R_o, α_o). Clearly, from Step 1, it now follows that R_o is a valuation ring.

Step 3. Let F, R be as in Step 1. Let $(F \supseteq) S \supseteq R$ be an integral extension. Then α extends to $\tilde{\alpha}: S \to K$.

Proof. Let $0 \neq x \in S$. Then x satisfies :

$$x^n + a_1 x^{n-1} + \ldots + a_n = 0 \qquad a_1, \ldots, a_n \in R$$

i.e.
$$x + a_1 + \frac{a_2}{x} + \ldots + \frac{a_n}{x^{n-1}} = 0$$

Hence
$$x \in R[x^{-1}].$$

Now, let R_o be as in Step 2. Then either $x \in R_o$ or $x^{-1} \in R_o$. If $x^{-1} \in R_o$, then $R[x^{-1}] \subseteq R_o$ and hence $x \in R_o$. In any case, $x \in R_o$, so that $S \subseteq R_o$. Hence α extends to $\tilde{\alpha} : S \longrightarrow K$.

Step 4. Deduction of the lemma from steps 1 to 3. Let $A \supseteq B$ be integral domains. $A = B[x_1, .., x_n]$. $0 \neq f \in A$ be given. We prove the lemma by induction on n. Assume the lemma proved for $n = 1$. Let $B^1 = B[x_1, \ldots, x_{n-1}]$, so that $A = B^1[x_n]$. Then $\exists\, 0 \neq g' \in B'$ with the required property. Now given this $g' \neq 0$, by induction hypothesis, $\exists\, 0 \neq g \in B$ with the required property. So let $n = 1$, $A = B[x]$; $0 \neq f \in A$ given.

Case (i). x is transcendental over B. Let $f = b_0 + b_1 x + \ldots + b_m x^m$ with $b_m \neq 0$. Choose $g = b_m$ (In fact any $b_i \neq 0$ will do). If $\alpha : B \longrightarrow K, \alpha(g) \neq 0$ then $\sum_{i=0}^{m} \alpha(b_i) X^i$ is not identically zero. Hence choose $\lambda \in K$ with $\sum_{i=0}^{m} \alpha(b_i). \lambda^i \neq 0$. Define $\tilde{\alpha}$ by $\tilde{\alpha}(x) = \lambda$. Then $\tilde{\alpha}$ is the required extension.

Case (ii). x satisfies a polynomial in $B[X]$. Consider F = quotient field of A. It follows that $B[x]$ is algebraic over B. Hence f^{-1} is also algebraic over B. Now choose $g \neq 0$, $g \in B$ such that gx and gf^{-1} are integral over B (such g exists). Let $\alpha : B \longrightarrow K$ be a homomorphism such that $\alpha(g) \neq 0$. It follows from Step 1 that α extends to $B[g^{-1}]$. Now from Step 3, α extends to $B[g^{-1}, gx, gf^{-1}]$. But then $B[g^{-1}, gx, gf^{-1}] \supseteq B[x] = A$. Hence α extends to A. Further, $f^{-1} \in B[g^{-1}, gx, gf^{-1}]$. Hence $\alpha(f) \neq 0$. This proves the lemma.

The lemma mentioned above has some fine applications.

Corollary to lemma 1. Every proper ideal I of $k[X_1,\dots,X_n]$ has a zero in k^n.

Proof. I can be assumed to be maximal. Consider $A = k[X_1,\dots,X_n]/I$, $\bar{x}_i = x_i \bmod I$, $B = k$, $f = 1$. Then $\exists\, g \neq 0$ with required properties. Consider $k \xrightarrow{Id} k$ and $Id(g) \neq 0$. This extends to $A \xrightarrow{\alpha} k$. Let $\alpha(\bar{X}_i) = a_i$. $(a_1,\dots,a_n)$ is clearly a zero of I.

Corollary 1 to the Proposition 1. Let G be a connected algebraic group acting on a variety V. Then every orbit is open in its closure.

Proof. Fix $v \in V$. Consider $G \xrightarrow{\alpha} V$ given by $\alpha(g) = g.v$. Then α is a morphism. G itself is an épais. Hence $\alpha(G)$ contains a open set in $\overline{\alpha(G)}$. Now $\alpha(G) = G.v$ = the orbit through v. Let U be a open set in $\overline{G.v}$ such that $g_0.v \in U \subseteq G.v$ for some $g_0 \in G$. Since G.v is invariant under G, it follows that $g.v \in gg_0^{-1}U \subseteq G.v$. Hence $G.v = \bigcup_{g \in G} gU$, which is open in $\overline{G.v}$.

Corollary 2. The closure $\overline{G.v}$ of an orbit is a union of G.v and other orbits of a smaller dimension. (We shall define this term presently).

Proof. Since $\overline{G.v}$ is invariant under the action of G, it follows that it is a union of orbits of which G.v is one. Clearly, the union of the other orbits is closed in $\overline{G.v}$. We define, for an irreducible variety V, the dimension to be equal to tr. $\deg._k k[V]$. For a reducible variety, we take the maximum dimension of any irreducible component. It can be proved that if $V \supsetneq W$; V irreducible, W closed, then $\dim V \gneq \dim W$. (Thus the maximum condition, as well as the minimum condition, on closed irreducible sets is satisfied).

It now follows that $\overline{G.v}$ is a union of $G.v$ and other orbits of a strictly smaller dimension.

<u>Corollary 3.</u> Orbits of minimal dimension are closed. Hence closed orbits exist.

The proof is obvious from corollary 2.

<u>Remarks.</u> (1) The corollaries 1, 2, 3 remain true if we remove the assumption of connectedness on G (exercise).

Consider the particular case when G acts on itself by conjugation, i.e. $x(y) = {}^{x}y = xyx^{-1}$. Then the orbits are the conjugacy classes. Thus we have:

<u>Corollary 4.</u> Every conjugacy class of an algebraic group is open in its closure and its closure is a union of the class and classes of strictly smaller dimensions.

<u>Proposition 2.</u> Let $\alpha : G \longrightarrow G'$ be a morphism of algebraic groups.

(a) $\alpha(G)$ is closed in G'.

(b) $\alpha(G^o) = \alpha(G)^o$.

(c) $\dim G = \dim \operatorname{Ker} \alpha + \dim \operatorname{Im} \alpha$.

<u>Proof.</u> (a) Let G act on G' via α i.e. $x(g') = \alpha(x).g'$. Then the orbits are the various cosets of $\alpha(G)$ in G'. Hence the orbits are <u>all</u> isomorphic to each other. Hence the orbits have the same dimension, which is obviously the minimum dimension. Hence by corollary 3, the orbits are closed. In particular, $\alpha(G) = \alpha(G).1$ is closed.

(b) Consider $\alpha: G \longrightarrow \alpha(G)$. $\alpha(G)$ is an algebraic group in its own right. $\alpha(G^o)$ being irreducible $\subseteq \alpha(G)^o$. But then $\alpha(G^o)$ is a closed subgroup having finite index in $\alpha(G)$. Hence $\alpha(G^o) \supseteq \alpha(G)^o$.

(c) This follows at once from a general elementary fact about morphisms, whose proof we shall omit, viz.

<u>Lemma 2.</u> Let $f : U \longrightarrow V$ be a morphism of algebraic varieties with U irreducible and $f(U)$ dense in V (such a morphism is called a dominant morphism). Then $\dim f^{-1}(w) \geqslant \dim U - \dim V \ \forall w \in V$. The equality holds for w in some open dense subset of $f(U)$.
(The part dealing with equality is proved in the Appendix to 2.11 below).

<u>Remark.</u> An alternate proof for (a) (without using corollary 3) is as follows:

Let G^o act on G' as before. Then $S = \alpha(G^o)$, being orbit of 1, is open in its closure. But then S is irreducible, hence so is $\overline{S}$. Hence for any $x \in \overline{S}$, xS^{-1} and S intersect. Hence $x \in S.S \subseteq S$. Thus $\overline{S} = S$. Hence $\alpha(G^o)$ is closed. Now $\alpha(G) = \bigcup_{i=1}^{n} \alpha(h_i).\alpha(G^o)$, where $G = \bigcup_{i=1}^{n} h_i G^o$ is the coset decomposition. Hence $\alpha(G)$ is closed also.

<u>Remark.</u> If k above is not algebraically closed, then essentially all of the results above do not hold, as the example $f : \mathbb{R}^* \longrightarrow \mathbb{R}^*$, $f(x) = x^2$, shows.

Chapter II

First Part: Jordan decompositions, unipotent and diagonalizable groups

2.1. Definitions and preliminary results. Let V be a finite dimensional vector space over k. (k will be algebraically closed unless stated otherwise).

Definition. An endomorphism X on V is said to be semisimple if it is diagonalizable (i.e. the eigenvectors of X span V).

It can be seen that X is semisimple iff the minimal polynomial of X has distinct roots iff $k[X]$ as a k-algebra is semisimple.

Definition. An endomorphism X on V is said to be nilpotent if $X^n = 0$ for some integer $n \geq 1$.

It can be seen that X is nilpotent iff all the eigenvalues of X are equal to zero.

Definition. An endomorphism X on V is said to be unipotent if $X-I$ is nilpotent.

It can be seen that X is unipotent iff all the eigenvalues of X are equal to 1.

From the above, it follows immediately that the restriction of a semisimple (respectively nilpotent, unipotent) endomorphism to an invariant subspace is again semisimple (respectively nilpotent, unipotent) and similarly for quotients.

It can also be seen that an endomorphism which is semisimple and nilpotent must be identically zero.

Lemma. Any commuting set of endomorphisms of a finite dimensional vector space V can be put simultaneously in an upper-triangular form in such a way that the semisimple elements are diagonal.

Proof. Let $\mathfrak{S}$ be a commuting set of endomorphisms. We use induction on the dimension of V. If $\dim V \leq 1$, every endomorphism is scalar and the lemma is trivially true. Assume the lemma for spaces W with $\dim W < \dim V$.

Case (i) $\mathfrak{S}$ contains a non-scalar semi-simple endomorphism A. Let $V = \dot{\sum\limits_{\alpha \in k}} V_\alpha$, where V_α is the eigenspace of A corresponding to $\alpha \in k$. It follows that for $\alpha \in k$, $\dim V_\alpha < \dim V$. Also, $S(V_\alpha) \subseteq V_\alpha$ $\forall S \in \mathfrak{S}$; since S and A commute. Hence by induction, the lemma is true for V_α $\forall \alpha \in k$. Clearly, the lemma is proved for V in this case.

Case (ii) $\mathfrak{S}$ does not have a non-scalar semisimple element. If $\mathfrak{S}$ consists of scalar elements only, then the lemma is obvious. If not, choose $A \in \mathfrak{S}$ such that A is not a scalar-element. Since k is algebraically closed, A has eigenvalues in k. Let α be one such value. Let V_α be the corresponding eigenspace; then $\{0\} \neq V_\alpha \neq V$. Also, $S(V_\alpha) \subseteq V_\alpha$ $\forall S \in \mathfrak{S}$. Hence $S \in \mathfrak{S}$ acts on V/V_α also. Now, by induction, the lemma is true for V_α and V/V_α. Clearly, the lemma is true for V also. Hence by induction, the lemma is true for all vector spaces (finite dimensional).

Corollary. For a set $\mathfrak{S}$ of semi-simple endomorphisms, the elements of S commute iff they can be simultaneously diagonalized.

2.2. Jordan decomposition for an endomorphism.

Proposition 1. Let V be an n-dimensional vector space over an algebraically closed field k. Let $X \in \operatorname{End} V$ (the set of endomorphisms of V). Then there exist endomorphisms $S, N \in \operatorname{End} V$ with the properties:

$X = S + N$; S is semi-simple, N is nilpotent and S and N commute.

Further, S, N are uniquely determined by the above conditions. These uniquely determined endomorphisms are polynomials (without constant term) in X (and are called the semisimple and nilpotent parts of X).

Proof. Consider the minimal polynomial $f(T)$ of $X, f(T) = \prod_{\alpha \in k} (T-\alpha)^{n_\alpha}$. It follows that α, involved above, is an eigenvalue of X. Let $V^\alpha = \{v \in V/(X-\alpha)^m(v) = 0$ for some integer $m\}$. Then it can be proved easily that: $V = \dot{\sum} V^\alpha$; $V^\alpha = \operatorname{Ker}(X-\alpha)^{n_\alpha}$; and $(T-\alpha)^{n_\alpha}$ is the minimal polynomial of X on V^α. Hence α is the only eigenvalue of X on V^α. Define S on V^α to be the scalar-multiplication by α. This defines S on V. Clearly, S is semisimple, S and X commute and (X-S) is nilpotent on each V^α and hence on V. Thus $X = S + (X-S) = S + N$ satisfies the required properties. Again, $\{(T-\alpha)^{n_\alpha}\}$ are coprime. Hence by the Chinese Remainder Theorem, there exists a polynomial p(T) such that

$$p(T) \equiv \alpha \ (\operatorname{mod}\ (T-\alpha)^{n_\alpha}) \text{ and } p(T) \equiv 0 \ (\operatorname{mod} T) \text{ if } V^0 = \{0\}.$$

On V^α, $(X-\alpha)^{n_\alpha} = 0$. Hence $p(X) = \alpha$ on V^α. Hence $p(X) = S$. It clearly follows that p(T) does not have constant term ($p(T) \equiv 0 \bmod T^n$ for some $n \geq 1$). Hence S, N are polynomials without constant term in X. If $X = S' + N'$ is another decomposition with the said properties, then S' commutes with X and hence with S and N (being polynomials in X). Similarly N' commutes with S and N. Also, $X = S + N = S' + N'$. Hence $S' - S = N - N'$. Now sums, products of commuting semisimple (respectively nilpotent) endomorphisms are semisimple (respectively nilpotent), e.g. by the above lemma. Hence $S' - S = N - N'$ is semisimple and nilpotent and hence equal to 0. Thus $S = S'$, $N = N'$. Hence the proposition.

Remark. If X is invertible, so is S (having the same eigenvalues). And S^{-1} is a polynomial in X.

For. Let p(T) be the polynomial such that $p(T) \equiv \alpha \pmod{(T-\alpha)^{n_\alpha}}$; α an eigenvalue of X. Then $p(X) = S$. But for X invertible, $\alpha \neq 0$. Hence p(T) and $f(T) = \prod (T - \alpha)^{n_\alpha}$ are coprime. Hence $\exists$ q(T), r(T) such that $1 = p(T).\, q(T) + r(T).\, f(T)$. Hence $Id = p(X).q(X)$; since $f(X) = 0$. Hence $S^{-1} = q(X)$. Thus S^{-1} is a polynomial in X.

Proposition 2. Any automorphism x of a finite dimensional vector-space can be uniquely expressed as a product of a semisimple and a unipotent automorphism which commute with each other.

i.e. $x = s.u$; s semisimple, u unipotent; s and u commute. Such s and u are unique.

Further, the uniquely determined automorphisms s and u are polynomials (without constant term) in X. These are called the semisimple and unipotent parts of x.

Proof. Let $x = S + N$ be the decomposition as in proposition 1. Since x is invertible, so is S. Let $s = S$ and $u = I + S^{-1}N$. Then $x = s.u$, s and u commute and u is unipotent. Further, the uniqueness follows from that in proposition 1 since given $x = s.u$, $S = s$ and $N = s.\,(u - I)$ give the decomposition $x = S + N$. From the remark to proposition 1, it follows that s,u are polynomials (without constant term) in X.

<u>Note.</u> The above notions can be defined, with a modification, for an arbitrary field K (not necessarily algebraically closed). e.g. we define an endomorphism in End V to be semisimple if the corresponding endomorphism on $V \otimes_K \overline{K}$ is semisimple. The above two propositions will go through, but we cannot assert whether the semisimple and nilpotent (or unipotent) parts are in End V. However, if K is perfect, then these parts do belong to End V.

<u>For.</u> Let $\sigma \in \mathrm{Gal}(\overline{K}/K)$. Let $X = S + N$ be the decomposition of $X \in \mathrm{End}\ V$. Then $\sigma X = \sigma S + \sigma N$, with σS semisimple, σN nilpotent and $\sigma S \cdot \sigma N = \sigma N \cdot \sigma S$. Also, $\sigma X = X$. Hence by uniqueness, $\sigma S = S$ and $\sigma N = N$. Since K is perfect, $S \in \mathrm{End}\ V$ as $\sigma S = S\ \forall \sigma \in \mathrm{Gal}(\overline{K}/K)$. (This follows immediately from Galois theory). Similarly $N \in \mathrm{End}\ V$.

This is not true in the case of a field which is not perfect. The following example brings out this fact. Let k_o be any field of characteristic $p \neq 0$. Let T be transcendental over k_o. Let $k = k_o(T)$. Then k is not perfect.

Let $V = k^p$. Consider the endomorphism on V given by : $X = \begin{bmatrix} 0 & 1 & & \\ & 0 & 1 & \\ & & 0 & 1 \\ T & & & 0 \end{bmatrix}_{p \times p}$.

Then the semisimple and nilpotent parts do not belong to End V, as the reader may check.

2.3. <u>Jordan decomposition for an endomorphism (continued).</u> We now partially extend this idea of decomposition in the case of infinite dimensional spaces.

Let V be a vector space (not necessarily finite dimensional).

Definition. $X \in \operatorname{End} V$ is called locally finite if $V = \sum_{\lambda \in \Lambda} V_\lambda$ where each V_λ is a finite dimensional subspace invariant under X; or, in other words, if each $v \in V$ is contained in a finite dimensional subspace invariant under X.

We know that for an infinite dimensional space, an endomorphism may be injective without being surjective. However, we have:

Lemma. Let $X \in \operatorname{End} V$ be locally finite. Then the following statements are equivalent.

(1) X is an automorphism (i.e. X is injective and surjective).

(2) All the eigenvalues of X are non-zero (i.e. X is injective).

Proof. (1) $\Longrightarrow$ (2) is clear.

(2) $\Longrightarrow$ (1). Let X be injective. Consider $V = \sum_{\lambda \in \Lambda} V_\lambda$ such that $X(V_\lambda) \subseteq V_\lambda$ for each λ with each V_λ finite dimensional. Now X/V_λ is injective and hence surjective. Thus X itself is surjective.

Note. It follows that in the above case, the inverse Y of X preserves the same finite dimensional subspaces as X and hence is locally finite.

Proposition 1. Let $X \in \operatorname{End} V$ be locally finite. Then there exist endomorphisms $S, N \in \operatorname{End} V$, such that (1) $X = S + N$ (2) S and N commute (3) If V' is a finite dimensional subspace invariant under X, then $S(V') \subseteq V'$, $N(V') \subseteq V'$ and $X/_{V'} = S/_{V'} + N/_{V'}$ is the decomposition of $X/_{V'}$ defined in the previous section.

Proof. Let $\left\{V_\lambda, \lambda \in \Lambda\right\}$ be the set of all finite-dimensional X-invariant subspaces of V. Since X is locally finite, $V = \sum_{\lambda \in \Lambda} V_\lambda$. Consider $X_\lambda = X/V_\lambda$.

Then by proposition 1 of the previous section, $X_\lambda = S_\lambda + N_\lambda$; $S_\lambda, N_\lambda \in \mathrm{End}\, V_\lambda$; S_λ semisimple, N_λ nilpotent; S_λ and N_λ commute. Define $S \in \mathrm{End}\, V$ by $S/_{V_\lambda} = S_\lambda \ \forall \lambda \in \Lambda$. Define $N \in \mathrm{End}\, V$ by $N/_{V_\lambda} = N_\lambda \ \forall \lambda \in \Lambda$.

Claim: S and N are well-defined.

Consider V_λ and V_μ. On $V_\lambda \cap V_\mu = W$, which is invariant under X,

$$\left.\begin{array}{l} X/_W = S_\lambda/_W + N_\lambda/_W \\ X/_W = S_\mu/_W + N_\mu/_W \end{array}\right\} \begin{array}{l}\text{are decompositions into semisimple} \\ \text{and nilpotent parts.}\end{array}$$

Hence by uniqueness of such a decomposition, $S_\lambda/_W = S_\mu/_W$ and $N_\lambda/_W = N_\mu/_W$. Hence S, N are well-defined. Since $S_\lambda . N_\lambda = N_\lambda . S_\lambda \ \forall \lambda \in \Lambda$, $S.N = N.S$. Also $X = S + N$. This proves the proposition.

Note that S and N defined above are locally finite. Further, S is semisimple (i.e. V has a basis consisting of eigenvectors of S). N is locally nilpotent (i.e. N is nilpotent on every finite-dimensional invariant space).

Proposition 2. Let $x \in \mathrm{Aut}\, V$ be locally finite. Then there exist automorphisms s, $u \in \mathrm{Aut}\, V$ such that (1) $x = s.u$, (2) s is semisimple, u is locally unipotent (see (4)), (3) s and u commute, (4) If V' is a finite dimensional subspace of V with $x(V') \subseteq V'$ then $s(V') \subseteq V'$, $u(V') \subseteq V'$ and $x/_{V'} = s/_{V'} . u/_{V'}$ is the decomposition for $x/_{V'}$.

The proposition is deduced from the previous proposition just as proposition 2 of 2.2 is deduced from proposition 1 of 2.2. (The only thing to be noted is that s has all of its eigenvalues non-zero and hence, in view of the lemma above, is an automorphism).

Remark. The decompositions here and in the previous sections will be called Jordan decompositions.

2.4. Jordan decomposition of an element x of an algebraic group G. Let G be an algebraic group and $A = k[G]$. In view of the lemma of 1.8, ρ_x^* and λ_x^* are locally finite $\forall\, x \in G$.

Definitions. (i) An element $s \in G$ is said to be semisimple if $\rho_s^* : A \longrightarrow A$ is so.

(ii) An element $s \in G$ is said to be unipotent if $\rho_s^* : A \longrightarrow A$ is so. (i.e. locally all the eigenvalues of ρ_s^* are equal to 1).

Note. s is semisimple (respectively unipotent) iff s^{-1} is so. Again, $\lambda_x = i \circ \rho_{x^{-1}} \circ i$. Hence $\lambda_x^* = i^* \circ \rho_{x^{-1}}^* \circ i^*$. Thus, ρ_x^* is semisimple iff λ_x^* is so.

Also, ρ_x^* is unipotent iff λ_x^* is so.

In other words, x is semisimple iff λ_x^* is semisimple.

x is unipotent if λ_x^* is unipotent.

Proposition 1. Let $x \in G$. Then there exist elements $y, z \in G$ such that $x = y.z = z.y$; y is semisimple, z is unipotent. Such elements are unique and are called the semisimple and unipotent parts of x. They are denoted by x_s and x_u respectively. $x = x_s . x_u$ is called the Jordan decomposition of x.

Proof. Consider $\rho_x^* : A \longrightarrow A$. Hence by proposition 2 of 2.3, there exist automorphisms s and u of A as a linear space such that

$$\rho_x^* = s.u = u.s \ ; \ s \text{ semisimple, } u \text{ unipotent.}$$

It follows that whenever σ commutes with ρ_x^*, σ commutes with s and u($\sigma \in$ End V). This is so since any $v \in V$ belongs to a finite dimensional X-invariant space and s is a polynomial of ρ_x^* on such a space. Hence $\sigma \cdot s(v) = s \cdot \sigma(v)$. Similarly, σ commutes with u. In particular, s and u commute with λ_t^* for every $t \in G$, since ρ_x^* clearly does so. ——— (1)

We now prove a lemma under a general set up.

Lemma 1. Let A be an algebra over k which is not necessarily commutative or associative. Let σ be a k-algebra automorphism of A which is locally finite. Let $\sigma = s.u$ be the Jordan decomposition of σ. Then s and u are also k-algebra automorphisms of A.

Proof. One need only prove that s is an algebra homomorphism. Since σ is locally finite, $A = \bigoplus_{\alpha \in k} A^{\alpha}$; where $A^{\alpha} = \{f \in A / (\sigma - \alpha)^n(f) = 0$ for some integer $n\}$. We also have : s is just multiplication by α on A^{α}.

Claim: $A^{\alpha} . A^{\beta} \subseteq A^{\alpha \cdot \beta}$. This clearly shows that $s(f.g) = s(f).s(g) \ \forall f \in A^{\alpha}$, $g \in A^{\beta}$ and hence $\forall f, g \in A$. For $f, g \in A$, the following identity can be proved by induction:-

$$(\sigma - \alpha.\beta)^n(fg) = \sum_{i=0}^{n} \binom{n}{i} . (\sigma - \alpha)^i (\alpha^{n-i} f) . \sigma^i (\sigma - \beta)^{n-i}(g).$$

If $f \in A^{\alpha}$, $g \in A^{\beta}$, then for large enough n, $(\sigma - \alpha.\beta)^n (fg) = 0$ thus $f.g \in A^{\alpha \cdot \beta}$. Hence the claim and the lemma.

The lemma immediately gives: If $\rho_x^* = s.u$ (as above), then s and u are k-algebra automorphisms of A. ——— (2)

Next we prove : Any k-algebra automorphism (or endomorphism) of A which commutes with λ_G^* is of the form ρ_w^* for unique $w \in G$. ——— (3)

Consider $x \in G$, $f \in A$. Then $\sigma \circ \lambda_x^* = \lambda_x^* \circ \sigma$. Hence $\sigma(\lambda_x^*(f)) = \lambda_x^*(\sigma(f))$. Consider the evaluation e_1 at the identity of G.

We have $$e_1(\sigma(\lambda_x^* f)) = (e_1 \circ \lambda_x^*)(\sigma(f))$$

i.e. $$(e_1 \circ \sigma)(\lambda_x^* f) = (e_1 \circ \lambda_x^*)(\sigma(f)).$$

Now, $e_1 \circ \sigma : A \longrightarrow k$, a k-algebra homomorphism. Hence $\exists\, w \in G$ such that $e_1 \circ \sigma = e_w$ by property (3) of varieties. Thus, $e_w(\lambda_x^* f) = e_x(\sigma(f))$, since $e_1 \circ \lambda_x^* = e_x$. Now, $e_w \circ \lambda_x^* = e_{xw} = e_x \circ \rho_w^*$. Thus $e_x(\rho_w^*(f)) = e_x(\sigma(f))$. Since $x \in G$ and $f \in A$ are arbitrary, $\rho_w^* = \sigma$. Now uniqueness of w follows immediately. <u>For</u>: $\rho_w^* = \rho_{w'}^*$ implies $e_1(\rho_w^*(f)) = e_1(\rho_{w'}^*(f))$ $\forall f \in A$. Hence $e_w(f) = e_{w'}(f)$ $\forall f \in A$ implies $w = w'$.

It follows immediately, from (1), (2) and (3), that $\rho_x^* = \rho_y^* \circ \rho_z^*$ for <u>unique</u> y, semisimple in G and z, unipotent in G. Hence $x = y.z$ by (3) again, since $\rho_y^* \circ \rho_z^* = \rho_{yz}^*$. Clearly y and z commute. This proves the proposition.

<u>Note</u>. (1) It follows immediately that for λ_x^*, which is also locally finite, $\lambda_x^* = \lambda_{x_s}^* \circ \lambda_{x_u}^*$ is the Jordan decomposition.

(2) For $g \in G$, ρ_g^* commutes with ρ_x^* iff it does so with $\rho_{x_s}^*$ and $\rho_{x_u}^*$ since these are polynomials in ρ_x^* locally. Hence $Z_G(x) = Z_G(x_s) \cap Z_G(x_u)$.

To complete the development, two more points remain to be proved.

<u>Proposition 2</u>. If G is a closed subgroup of GL(V) and $x \in G$, then the two Jordan decompositions of x, one as automorphism of V and the other as an

element of G, are one and the same.

Proposition 3. The Jordan decomposition is preserved by morphisms of algebraic groups.

Both these propositions follow from:

Lemma 2. Let $\rho : G \longrightarrow GL(V)$ be a representation (i.e. a morphism into some GL(V)) and $x \in G$. Then $\rho(x) = \rho(x_s).\rho(x_u)$ is the Jordan decomposition of $\rho(x)$ as an automorphism of V.

To get proposition 2, we apply this lemma to the injection: $G \hookrightarrow GL(V)$. To get proposition 3, assuming $\alpha : G \longrightarrow G'$ is the given morphism, we imbed G' in some GL(V) and then apply proposition 2 to the resulting representations of G and of G'. Lemma 2 itself follows from:

Lemma 3. Every representation of G is isomorphic to one in which G acts via right-translations on a subspace of A^n for some n, $A = k[G]$.

For: In view of the definitions and the fact that the Jordan decomposition on vector spaces is preserved by direct sums and restrictions to subspaces, lemma 3 clearly implies lemma 2.

Proof of lemma 3. Let $\rho : G \longrightarrow GL(V)$ be the given representation. Let $\{v_1^*, v_2^*, \dots, v_n^*\}$ be a basis of the dual space of V. Consider the map $\phi : V \longrightarrow A^n$ given by : $\phi(V) = (C_{v_1 v_1^*}, \dots, C_{v_1 v_n^*})$; where, as usual, $C_{v_1 v^*}$ stands for the matrix coefficient $x \rightsquigarrow v^*(\rho(x)(v_1))$. Now, it can be easily verified that ϕ yields a G-module isomorphism of V with $\phi(V)$. This proves the lemma.

Remark. The Jordan decomposition may be carried over from elements to

Abelian groups:

If G is an Abelian affine group, then $G = G_s . G_u$, a direct product (in the sense of algebraic groups) of closed subgroups.

The reader may wish to prove this using the lemma of 2.1.

2.5. Kolchin's Theorem. We continue with an important theorem about unipotent groups.

Theorem (Kolchin). Let G be a subgroup of Aut V (for a finite dimensional space V) consisting of unipotent elements. Then the elements of G can be simultaneously put in uppertriangular form (i.e. G fixes a flag: $0 = V_o \subsetneq V_1 \subsetneq \ldots \subsetneq V_n = V$ such that V_i is of codimension 1 in V_{i+1}, $0 \leq i \leq n-1$).

Proof. For the proof, we make use of lemma :

Lemma 1 (Burnside). Let S be a semigroup of endomorphisms of V (finite dim.) such that V is a simple S-module (i.e. the action of S on V is irreducible). Then S contains a basis of End V.

From this lemma, the following lemma follows:

Lemma 2 (Burnside). Let V be an n-dimensional space$/_k$, S be a semigroup of endomorphisms of V such that V is simple S-module. Let the elements of S have only r different traces (i.e. let the cardinality of the set: $\{\text{tr } s, s \in S\}$ be r). Then S itself is finite and the number of elements $\leq r^{n^2}$.

Proof. By Lemma 1, S contains elements $y_1, \dots, y_{n^2}$ which form a basis of End V. Consider the map $S \xrightarrow{\phi} k^{n^2}$ given by :

$$\phi(S) = (\mathrm{tr}\, x.y_1, \dots, \mathrm{tr}\, x.y_{n^2})$$

Since $y_1, \dots, y_{n^2}$ is a basis of End V, it follows that ϕ is injective. But then $xy_i \in S, 1 \leq i \leq n^2$. Hence tr xy_i has r possible values. Hence S is finite and $|S|$ = number of elements in S $\leq r^{n^2}$.

From this, Kolchin's theorem follows immediately. Consider a composition series $V = V_r \underset{\neq}{\supset} V_{r-1} \underset{\neq}{\supset} \dots \underset{\neq}{\supset} V_0 = (0)$ for V with respect to G, i.e. $G(V_i) \subseteq V_i$ and V_i/V_{i-1} is a simple G-module $\forall\, 1 \leq i \leq r$. Let $\rho_i : G \rightarrow \mathrm{End}\, V_i/V_{i-1}$ is the corresponding representation. Elements of G are unipotent. Hence the action of G on V_i/V_{i-1} is also unipotent and there is only one trace-value on $\rho_i(G)$. Hence by lemma 2 above, $|\rho_i(G)| \leq 1$. Hence $\rho_i(G)$ = Identity. Hence every subspace of V_i/V_{i-1} is G-invariant. Since V_i/V_{i-1} is simple, it follows that V_i/V_{i-1} is of dimension 1. This proves the required result.

Remarks. (1) The above theorem holds for arbitrary fields (not necessarily algebraically closed) (check this).

(2) Arguing as in Lemma 2 one sees that a subgroup of $GL_n(k)$ with just r conjugacy classes is finite, of order at most r^{n^2}. By modifying the proof somewhat, one can prove that if char k = 0, S is a subgroup of Aut V without unipotent elements and with r traces only, then $|S| \leq r^{n^2}$, n = dimension of V. One can then easily deduce that over fields of characteristic 0, every torsion subgroup of Aut V with the elements of bounded orders is finite and that every torsion subgroup of $GL_n(\mathbb{Z})$ (i.e. matrices with integral coefficients and having inverses with integral coefficients also) is finite, of order $\leq (2n+1)^{n^2}$.

These results all go back to Burnside.

(3) Kolchin's theorem incidentally proves that every unipotent group is nilpotent since the group of upper triangular unipotent matrices is so (check this).

This result has an interesting consequence:

Proposition. Let G be a unipotent algebraic group (i.e. an algebraic group consisting of unipotent elements). Let G act on an affine variety V. Then every orbit is closed.

Proof. Let $\alpha : G \times V \longrightarrow V$ be the action. Write $\alpha(x, v) = x.v$. Let $\alpha^* : k[V] \longrightarrow k[G] \otimes k[V]$ be the co-morphism. For $x \in G$, define $x^* : k[V] \longrightarrow k[V]$ by :

$$(x^*f)(v) = f(\alpha(x^{-1}, v)) = f(x^{-1}.v).$$

Thus x^* is a k-algebra homomorphism of $k[V]$. Also, $x^* \circ y^* = (xy)^*$. Thus we get a map : $G \longrightarrow \text{Aut}\, k[V]$ which is a group-homomorphism. Let $f \in k[V]$. Then an argument similar to the one in the lemma of 1.7 gives: The space W(f) spanned by $\{x^*f, x \in G\}$ is finite dimensional and invariant under $G^* = \{x^*, x \in G\}$. Hence it follows that G acts locally finitely. Further, it can be seen, as in the proof of the lemma of 1.9, that the map $\phi : G \longrightarrow GL(W(f))$ given by $\phi(x) = x^*/_{W(f)}$ is a morphism of algebraic groups. Hence by proposition 2 of 2.4, we get: $x^*/_{W(f)}$ is unipotent $\forall x \in G$. Since $k[V] = \sum_{f \in k[V]} W(f)$, it follows that x^* is locally unipotent $\forall x \in G$. Consider an orbit O of V under G. If possible, let O be not closed. Consider $\bar{O}$. Then by corollaries 1 and 2 to proposition 1 of 1.2, it follows that O is open in $\bar{O}$ and $\bar{O} - O$ is a union of orbits (of smaller dimensions). Since $\bar{O} - O$ is a proper closed subset of $\bar{O}$, there exists $f \in k[\bar{O}]$ such that $f \equiv 0$ on $\bar{O} - O$ and $f \neq 0$ on O. (*)

Since $\bar{O} - O$ is a union of orbits, it follows that $x^*f(x \in G)$ has similar properties. Now $\{x^*/_{W(f)}, x \in G\}$ is a unipotent group. Hence by Kolchin's theorem, it has a common eigenvector f_0. Since $f_0 \neq 0$ and every element of W(f) is zero on $\bar{O} - O$, it follows that f_0 also has the above mentioned property (*). Now, $x^*(f_0) = f_0 \; \forall x \in G$. Hence $f_0(x^{-1}.v) = f_0(v) \; \forall x \in G, v$ being fixed. Hence f_0 is constant λ on O. But then the set $\{x \mid f_0(x) = \lambda\}$ is closed in $\bar{O}$ and contains O. Hence it is the whole of $\bar{O}$. Thus $f_0 = \lambda$ on $\bar{O}$. But f_0 is already 0 on $\bar{O} - O$ and $\bar{O} - O$ is non-empty by assumption. Hence $f_0 = 0$, a contradiction to the fact : $f_0 \neq 0$. Hence $\bar{O} - O$ is empty i.e. O is closed. This proves the proposition.

Corollary. Every conjugacy class of a unipotent algebraic group is closed.

The most important example of a unipotent algebraic group, incidentally, is the additive group G_a defined by $G_a(k) = (k, k[X])$ with addition the group operation. This group may be seen to be unipotent either from the isomorphism $t \leftrightarrow \begin{bmatrix} 1 & t \\ 0 & 1 \end{bmatrix}$ $(t \in k)$ or else directly from the form of the right translations in terms of the basis $1, X, X^2, \ldots$ of $k[X]$.

2.6. Diagonalizable Groups.

Definition. An (affine) algebraic group is said to be diagonalizable if it is commutative and consists of semisimple elements.

The most important example is the multiplicative group $G_m(k)$ of k, equal to $(GL(k), k[X, X^{-1}])$.

Proposition 1. For an algebraic group G, the following statements are equivalent:

(a) G is diagonalizable.

(b) G is isomorphic to a closed subgroup of some D_n (i.e. of the group of diagonal matrices in GL_n) or, equivalently, of some GL_1^n.

(c) $k[G]$ is spanned, as a vector space, by the characters. (A character is a morphism of G into GL_1.)

Proof. (a) $\Longrightarrow$ (b). This is obvious from the proposition of 1.9, proposition 2 of 1.13 and proposition 3 of 2.4.

(b) $\Longrightarrow$ (c). For a closed subgroup of D_n, $X_{11}^{m_1} \dots X_{nn}^{m_n}$ is a character, $m_i \in \mathbb{Z}$, $1 \leq i \leq n$. (By X_{ii}, we mean the canonical function taking an element of D_n to the ith (diagonal) entry). Obviously, $k[G]$ consists of polynomials which are linear combinations of such characters. Hence the characters span $k[G]$.

(c) $\Longrightarrow$ (a). Let f be a character of G. Then $f(xy) = f(x).f(y)\ \forall x,y \in G$. Hence $\rho_y^* f = f(y).f\ \forall y \in G$. In other words, every character is an eigen-vector for ρ_y^*, $y \in G$. Since characters span $k[G]$, it follows that ρ_y^* is semisimple and any two ρ_y^*, ρ_z^* commute. Hence G is diagonalizable.

Definition. The characters of an algebraic group G form a group under pointwise multiplication. The group is called the character group of G and and is denoted by X(G) (or simply X). Clearly X(G) is abelian.

The character group plays an important role in the theory of diagonalizable groups. This will become clear as we proceed with the development.

We now prove a proposition for diagonalizable groups. (This proposition holds for arbitrary algebraic groups).

Proposition 2. For a diagonalizable group G, X(G) is finitely generated.

Proof. Since G is diagonalizable, k[G] is spanned by X(G). But then k[G] is finitely generated as k-algebra. Hence there exist characters $X_1,\dots,X_n$ which generates k[G] as k-algebra. Let H be the subgroup generated by $X_1,\dots,X_n$. Clearly elements of H are of the form: $X_1^{r_1}\dots X_n^{r_n}$ with $r_i \in \mathbb{Z}$. Also, any $f \in k[G]$ is a linear combination of elements of H.

Claim. H = X(G). Let $X \in X(G)$. Then $X = \sum_{j=1}^{r} \lambda_j \cdot \eta_j$, $\eta_j \in H$. We may assume that the η_j's are all distinct. The proposition now follows from the following general lemma:

Lemma 1. Distinct characters of an (abstract) group H into k^* are linearly independent as k-valued functions on H.

Proof. Let, if possible, there exist relations between distinct characters. Let $\alpha_o + \sum_{i=1}^{r} \lambda_i \alpha_i = 0$, where α_i's are distinct characters of H into k^* and r is minimal with this property ($r \geq 1$). Choose $h_o \in H$ such that $\alpha_o(h_o) \neq \alpha_1(h_o)$. Consider

$$0 = \alpha_o(h_o \cdot h) + \sum_{i=1}^{r} \lambda_i \alpha_i(h_o \cdot h), \forall\, h \in H$$

$$= \alpha_o(h_o) \cdot \alpha_o(h) + \sum_{i=1}^{r} \lambda_i \cdot \alpha_i(h_o) \cdot \alpha_i(h).$$

Also, $$0 = \alpha_o(h_o) + \sum_{i=1}^{r} \lambda_i \alpha_i(h_o).$$

Hence $$0 = \sum_{i=1}^{r} (\alpha_i(h_o) - \alpha_o(h_o)) \cdot \alpha_i(h), \forall\, h \in H$$

and $\alpha_1(h_o) \neq \alpha_o(h_o)$. This contradicts the minimality of r. This proves the lemma.

Note. From this, it follows that for a diagonalizable group G, $X(G)$ is a basis of $k[G]$. It also follows, assuming H to be a closed subgroup of G, both diagonalizable, that every character of H can be extended to one of G, and that H is just the kernel of a set of characters on G. (Exercise: Prove these assertions.)

It can be easily verified that $X(G_1 \times G_2) = X(G_1) \times X(G_2)$ for diagonalizable groups G_1 and G_2.

Notation: Denote $p = p(k) = \begin{cases} \text{char. } k & \text{if char. } k \neq 0 \\ 1 & \text{if not.} \end{cases}$

Note. For an algebraic group G, $X(G)$ does not have p-torsion ($p = p(k)$ as defined above).

For: Let $X^p = 1$. Then $X^p(x) = 1, \forall\, x \in G$.

$$\Rightarrow (X(x) - 1)^p = 0 \quad \forall\, x \in G \Rightarrow \underline{\underline{X = 1.}}$$

Thus we have: For a diagonalizable group G, $X(G)$ is a finitely generated abelian group with torsion prime to p.

The converse is also true, viz.

Proposition 3. If X is a finitely generated abelian group with torsion prime to $p(= p(k))$, then there exists a diagonalizable group G such that $X(G) = X$.

Proof. A finitely generated abelian group is a (finite) direct product of cyclic groups. Since $X(G_1 \times G_2) = X(G_1) \times X(G_2)$, it follows that we need consider the following cases only.

Case (a). $X = \mathbb{Z}$. In this case, clearly $G = GL(1)$ is such that $X(G) = X$.

Case (b). $X = \mathbb{Z}/_{n\mathbb{Z}}$ with $(n,p) = 1$. In this case, we can take for G the closed subgroup of $GL(1)$, i.e. of k^*, consisting of the nth roots of 1. Since $(n,p) = 1$, this group is isomorphic to $\mathbb{Z}/_{n\mathbb{Z}}$. Hence $X(G)$ is also isomorphic to $\mathbb{Z}/_{n\mathbb{Z}}$ as can be verified at once. This proves the proposition.

We note that the group $\mathbb{Z}/_{n\mathbb{Z}}$ of case (b) is discrete.

Proposition 4. Let G, G' be diagonalizable groups. Then every homomorphism $\alpha^* : X(G') \longrightarrow X(G)$ comes naturally from a morphism $\alpha : G \longrightarrow G'$.

Proof. Since G' is diagonalizable, $X(G')$ is a basis of $k[G']$. Hence α^* extends to a k-algebra homomorphism of $k[G']$ into $k[G]$. This gives rise to a map $\alpha : G \longrightarrow G'$ given by: $e_{\alpha(u)} = e_u \circ \alpha^*$. Clearly α is a morphism of varieties. For $x, y \in G$ and $f \in X(G')$,

$$f(\alpha(xy)) = (\alpha^* f)(xy) = (\alpha^* f)(x).\ (\alpha^* f)(y),\ (\text{since } \alpha^* f \in X(G))$$

$$= f(\alpha(x))\ .\ f(\alpha(y)) = f(\alpha(x).\ \alpha(y)).$$

Since $X(G')$ spans $k[G']$, $\alpha(xy) = \alpha(x).\ \alpha(y)$. Hence α is a group homomorphism and hence a morphism. This proves the proposition.

Propositions 3 and 4 prove the following theorem.

Theorem. The correspondence $G \longrightarrow X(G)$ between diagonalizable groups and finitely generated abelian groups with torsion prime to p induces a fully faithful contravariant functor of categories.

Proposition 5. Let G be a diagonalizable group. Then the following statements are equivalent:

(a) G is connected.

(b) G is isomorphic to some GL_1^n.

(c) $X(G)$ is free, i.e. has no torsion.

Proof. Because of the theorem above, G is isomorphic to a group of the form:

$$GL_1^n \times \mathbb{Z}/_{n_1\mathbb{Z}} \times \dots \times \mathbb{Z}/_{n_r\mathbb{Z}} \quad \text{with } (n_i, p) = 1 \ \forall\, 1 \leq i \leq r.$$

and then $X(G) = \mathbb{Z}^n \times \mathbb{Z}/_{n_1\mathbb{Z}} \times \dots \times \mathbb{Z}/_{n_r\mathbb{Z}}$.

Clearly $GL_1^n = G^o$. Hence G is connected iff $r = 0$ iff $X(G) = \mathbb{Z}^n$. This proves the proposition.

Definition. A diagonalizable group G is said to be a torus if the above set of equivalent conditions holds for G.

As an immediate corollary, we get:

Corollary. Every diagonalizable group G is a direct product of a torus and a finite abelian group (with torsion prime to p). The torus is uniquely determined as the identity component of G.

Proposition 6. (a) The elements of finite order of a diagonalizable group G are dense. For a given integer n, there exists only a finite number of elements with order n.

(b) If k is not the algebraic closure of a finite field and G is a torus over k, then there exists an element $x \in G$ whose powers are dense in G.

Proof of part (a) follows immediately from the decomposition

$$G \xrightarrow{\sim} GL_1^n \times \mathbb{Z}/n_1\mathbb{Z} \times \dots \times \mathbb{Z}/n_r\mathbb{Z} .$$

(The Zariski topology on GL_1 is the cofinite topology and roots of unity are infinitely many in number).

The proof of part (b) (which we shall not use) may be found on page 208 of Borel's book.

We now prove a proposition which is very useful in later discussions.

Proposition 7. Let T be a torus and $\alpha_1, \dots, \alpha_r$ be linearly independent in X(T)(X(T) is a free $\mathbb{Z}$-module). Let $C_1, \dots, C_r \in k^*$ be arbitrarily given. Then $\exists\, t \in T$ such that $\alpha_i(t) = C_i \ \forall 1 \leq i \leq r$.

Proof. Consider the morphism $f : T \longrightarrow GL_1^r$, given by: $f(x) = (\alpha_1(x), \dots, \alpha_r(x))$. Since $\alpha_1, \dots, \alpha_r$ are linearly independent in X(T), monomials in $\alpha_1, \dots, \alpha_r$ are distinct elements of X(T) and hence are linearly independent as functions on T. It follows that $\alpha_1, \dots, \alpha_r$ are algebraically independent as functions on T. Hence f^* is injective so that f is dominant i.e. f(T) is dense in GL_1^r. But f is a morphism of groups and hence by proposition 2 of 1.13, f(T) is closed in GL_1^r. The proposition now follows.

2.7. Rigidity Theorem. We proceed to prove an important theorem.

Theorem. Let $\alpha : V \times H \longrightarrow H'$ be a morphism such that:

(1) H is an algebraic group in which the elements of finite order are dense.

(2) H' is an algebraic group which contains only finitely many elements of a given finite order.

(3) V is a connected variety.

(4) For a fixed $v \in V$, the map $\alpha_v : H \longrightarrow H'$, given by $\alpha_v(h) = \alpha(v,h)$, is a morphism of groups. Then $\alpha_v : H \longrightarrow H'$ is the same morphism $\forall v \in V$ (i.e. α factors through $p_2 : V \times H \longrightarrow H$).

Proof. Let $h \in H$ be of finite order. Consider $\alpha(V \times h)$ which consists of elements having orders which divide the order of h. Hence by (2), $\alpha(V \times h)$ is a finite set. But it is connected also, since V is connected. Hence $\alpha(V \times h)$ consists of a single element. Thus, whenever $h \in H$ is of finite order, $\alpha(v,h) = \alpha(v',h)\ \forall v, v' \in V$. For fixed $v, v' \in V$, the above is a 'polynomial' condition on H and hence defines a closed subset F of H. This closed subset contains the set of all elements with finite orders. Hence by (1), $F = H$, i.e. $\alpha(v,h) = \alpha(v',h)\ \forall h \in H$. Since $v, v' \in V$ are arbitrary, it follows that $\alpha(v,h) = \alpha(v',h)\ \forall v, v' \in V,\ h \in H$. Thus $\alpha_v = \alpha_{v'}\ \forall v, v' \in V$. This proves the proposition.

Corollary 1. Since conditions (1), (2) are satisfied in the case of diagonalizable groups H, H', the above proposition holds for these groups.

Corollary 2. Let H be a diagonalizable subgroup of an algebraic group G. Then we have the following:

(a) $N_G(H)^o = Z_G(H)^o$. Hence if G is connected and H is a diagonalizable normal subgroup then H is central.

(b) $N_G(H)/Z_G(H)$ is finite.

Proof. In the corollary 1, let $V = N_G(H)^o$, which is a connected variety. Take $H' = H$ and $\alpha : V \times H \longrightarrow H'$ to be the morphism $\alpha(v,h) = vhv^{-1}$. Hence vhv^{-1} is independent of v, by corollary 1. Taking $v = e$, $vhv^{-1} = e.h.e^{-1} = h$ $\forall v \in V$. In other words, $N_G(H)^o \subseteq Z_G(H)$. Hence $N_G(H)^o \subseteq Z_G(H)^o$. Thus $N_G(H)^o = Z_G(H)^o$. (Since $Z_G(H)^o \subseteq N_G(H)^o$ is always true.) Further, $N_G(H)/Z_G(H)$ is a homomorphic image of $N_G(H)/Z_G(H)^o = N_G(H)/N_G(H)^o$. The last mentioned group is finite by proposition 1 of 1.12. Hence the corollary is proved.

Proposition. (a) Let G be a diagonalizable algebraic group, acting on an affine variety V. Then only finitely many fixed point sets under the action of subsets of G occur in V. Also, only finitely many subgroups of G occur as stabilizers of subsets of V.

(b) If G is also connected (i.e. if G is a torus), then there exists an element $x \in G$ such that $x.v = v$ implies $y.v = v$ $\forall y \in G$ (i.e. $V_G = V_x$); in fact, for most $x \in G$ this is so.

Proof. G acts on $k[V]$ morphically, hence via semi-simple endomorphisms. Choose a finite set $\{f_1, \ldots, f_n\}$ of eigenvectors of the action of G which generates $k[V]$ as k-algebra. Since f_i is an eigenvector for each $x \in G$, there exists a character $X_i \in X(G)$ such that: $f_i(x.v) = X_i(x).f_i(v)$ $\forall x \in G, v \in V$. Now, $x.v = v$ iff $f_i(x.v) = f_i(v)$ $\forall i$ (since $\{f_1, \ldots, f_n\}$ generates $k[V]$) iff $f_i(v)(X_i(x) - 1) = 0$ $\forall 1 \leq i \leq n$. ——— I

(a) Let $W \subseteq V$ be the set of fixed points of a subset S of G. Let $J = \{i / X_i/_S \neq 1\}$. Then $W = \{v \in V / f_i(v) = 0 \ \forall i \in J\}$.

For: From I, $x.v = v \ \forall x \in S$ iff $f_i(v)(X_i(x) - 1) = 0 \ \forall x \in S, \forall i$ iff $f_i(v) = 0 \ \forall \ i \in J$. Thus W depends entirely on J which is a subset of $\{1, 2, \ldots, n\}$. Thus such W's are finitely many in number.

The statement about occurence of finitely many stabilizers can be proved in a similar way.

(b) If G is a torus, then G is irreducible as a variety. Consider $G_i = \{x \in G / X_i(x) \neq 1\}$. Then each G_i is an open set and $G_i \neq \emptyset$ whenever $X_i \neq 1$ on G. Let $J = \{i \mid G_i \neq \emptyset\}$. Since G is irreducible, $\bigcap_{i \in J} G_i \neq \emptyset$. So let $x \in \bigcap_{i \in J} G_i$.

Claim: $V_G = V_x$.

Now $V_G \subseteq V_x$ is always true. So let $v \in V$ such that $x.v = v$. Hence $f_i(v)(X_i(x) - 1) = 0 \ \forall i$. Now for $i \in J$, $X_i(x) \neq 1$ so that $f_i(v) = 0$. Hence for any $y \in G$, $f_i(v)(X_i(y) - 1) = 0 \ \forall i \in J$. But then for $i \notin J$, $X_i(y) = 1$. Hence $f_i(v)(X_i(y) - 1) = 0 \ \forall \ i \notin J$. Thus $y.v = v$ or $v \in V_G$. This proves the part (b).

Corollary. Let H be a diagonalizable subgroup of G.

(a) Only finitely many centralizers in G of subsets of H occur. Also, only finitely many centralizers in H of subsets of G occur.

(b) If H is connected, then $Z_G(H) = Z_G(x)$ for some $x \in H$; in fact, for most x.

Proof. Make H acts on G by conjugation (i.e. $h(g) = hgh^{-1} \ \forall h \in H, g \in G$). The Corollary now follows.

Second Part: Quotients and solvable groups

2.8. Solvable Groups. The basic result here is as follows:

Theorem 1 (Lie-Kolchin). A connected, solvable linear algebraic group fixes a flag (of the underlying space) i.e. can be put into an upper-triangular form.

A proof of this theorem will be given later (see 2.11). For an alternate proof, which could be given now, see Serre's book 'Lie Algebras and Lie Groups : L.A 5.11' or else the author's lectures on Chevalley groups.

Theorem 2. Let G be a connected solvable algebraic group.

(a) $G_u = \{g \in G \mid g \text{ unipotent}\}$ is a closed, connected normal subgroup of G containing $DG = [G, G]$; and hence the latter is nilpotent.

(b) If G is nilpotent, then $G_s = \{g \in G/g \text{ semisimple}\}$ is a (closed) torus and the direct product decomposition $G = G_s . G_u$ holds (i.e. the canonical map $m : G_s \times G_u \longrightarrow G$; $m(s,u) = s.u$ is an isomorphism of algebraic groups).

(c) The maximal tori of G are conjugate. If T is one of them, then $G = T.G_u$ is a semi-direct product. (The geometric requirement is that $T \times G_u \longrightarrow G$ is an isomorphism of varieties).

(d) If S is a subgroup of G, consisting of semisimple elements, then S can be imbedded in a torus. (Hence S is abelian).

(e) $N_G(S)$ is connected (S as in (d)) and hence is equal to $Z_G(S)$ (by the Rigidity Theorem).

Here the results important for our purposes have been underlined. A proof may be found in Borel's book (page 244). It uses theorem 1, induction on the length of the derived series, and the time-honoured method of averaging over finite groups which turns out to be applicable because the elements of finite order in a torus are dense.

Corollary. (a) Every unipotent element (in fact, subgroup) can be imbedded in a connected unipotent group.

(b) Every semisimple element (in fact, commutative subgroup of such elements) can be imbedded in a connected such group.

Remark. For nonsolvable connected groups (a) continues to hold, but (b) fails, as will be seen later.

2.9. Varieties in general. To continue, we have to consider varieties, more general than the affine ones. Roughly, a variety is a collection of a finite number of affine varieties, suitably patched together. More precisely, we have:

Definition. A variety is a topological space V with a finite cover $\{U_i\}_{1 \leq i \leq n}$ of open subsets satisfying the following properties:

(1) Each U_i is an affine variety.

(2) $U_i \cap U_j$ is a principal open set in both the affine varieties U_i and U_j and the identity map of $U_i \cap U_j$ is an isomorphism of the two affine structures on $U_i \cap U_j$ (obtained from U_i and U_j).

(3) The set of points $(x,x) \in U_i \times U_j$ $(x \in U_i \cap U_j)$ is closed there.

Note that an affine variety is a variety. Further U is open (resp. closed) in V iff $U \cap U_i$ is so in U_i (in the Zariski topology) for every i. A closed subset of a variety is a variety in a natural way, and so is an open one since an open subset of an affine variety is the union of a finite number of principal affine open subsets. (Check all of this.)

For a variety V, we write $k[V]$ for the algebra of rational functions on V that are defined everywhere. A function f is said to be defined at x if for some affine neighborhood U_x of x, $f = g/h$ with $g,h \in k[U_x]$ (in the old sense) and $h(x) \neq 0$. If V itself is affine, then $k[V]$ as just defined agrees with $k[V]$ in the old sense (by 1.13, cor. to lemma 1).

<u>Definition</u>. Let V, W be varieties. A map $f : V \to W$ is called a morphism if (1) f is continuous, (2) for all affine open sets $S \subseteq V$, $T \subseteq W$ with $f(S) \subseteq T$, the map $f : S \to T$ is a morphism of affine varieties

We can construct the product of two varieties by taking the products of the constituent affine varieties and then patching them together suitably. It and the resulting projections satisfy the universal property mentioned in 1.6 in the affine case. The condition (3) above can then be restated (check this):

(3') In $V \times V$ the diagonal is closed.

If we were using the product topology on $V \times V$ (which we aren't), this would say that the topology on V is Hausdorff (which it isn't).

Various of the properties of an affine variety continue to hold for an arbitrary variety, e.g. the decomposition into irreducible components. For further details see Mumford's book.

2.10. Complete varieties and Projective varieties.

Definition. A variety V is said to be complete if for every variety W, the projection map $p_2 : V \times W \longrightarrow W$ is closed.

Remarks. The affine line $\mathbb{A}^1 = (k, k[X])$ is not complete. In fact, as we shall see presently, any complete affine variety consists of finitely many points.

Proposition 1. (1) A closed subvariety of a complete variety is complete. The image of a complete variety under a morphism is closed and complete. Products of complete varieties are complete.

(2) A complete affine variety consists of finitely many points.

(3) A morphism from a connected complete variety to an affine variety is constant.

Proof. (1) Clearly a closed subvariety of a complete variety is complete. Let $f : V \longrightarrow W$ be a morphism of varieties with V complete.

Claim: $f(V)$ is a closed subvariety of W.

Consider $\Gamma = \{(v, f(v)) ; v \in V\} \subseteq V \times W$, the graph of f.

If we now prove that Γ is closed, then $f(V)$ will be closed in W, being the image of Γ under the map $p_2 : V \times W \longrightarrow W$. ($V$ is complete). The fact that Γ is closed, holds for any morphism $f : V \longrightarrow W$ (i.e. V may not be complete). For : The diagonal Δ of $W \times W$ is closed by (3') of 2.9 and Γ is the inverse image of Δ under the morphism $f \times Id : V \times W \rightarrow W \times W$.

Returning to the proof of the fact that $f(V)$ is complete, we see that $f(V)$ is a subvariety in its own right (being closed in W). For any variety T, consider the maps: $V \times T \xrightarrow{f \times Id} f(V) \times T \xrightarrow{p_2} T$. For a closed set $S \subseteq f(V) \times T$, $p_2(S) = p_2 \circ (f \times Id)((f \times Id)^{-1}(S))$ which is closed in T since V is complete. This proves the required result.

(2) Let W be a complete affine variety. Let A be its algebra of functions.

<u>Claim:</u> The set $f(W)$ is finite for every $f \in A$.

Consider the affine line $\mathbb{A}^1$ and the map $W \times \mathbb{A}^1 \xrightarrow[p_2]{} \mathbb{A}^1$. Let $S = \{(x, y) \in W \times \mathbb{A}^1 / f(x).y = 1\}$, $f \in A$. Clearly, S is closed in $W \times \mathbb{A}^1$. Also, $p_2(S)$ cannot be the whole set $\mathbb{A}^1$ (since $0 \notin p_2(S)$). But then W is complete and hence $p_2(S)$ is closed in $\mathbb{A}^1$. Hence $p_2(S)$ is a finite set in $\mathbb{A}^1$. It now follows that $f(W)$ is also finite.

From this claim, it follows immediately that W consists of finitely many points. (One can consider W as a subvariety of some k^n and then consider the coordinate functions X_i, $1 \leq i \leq n$).

<u>Note.</u> As an immediate deduction, one gets: A connected, complete, affine variety consists of a single element.

(3) Since the image of a connected variety is connected (as a set), (3) immediately follows from (1) and (2) above.

We now consider a very important class of complete varieties, viz. the projective varieties. We begin with:

<u>Projective Spaces</u>. Consider k^{n+1}. Let $\mathbb{P}^n$ be the set of all lines through 0 in k^{n+1}. One can easily see that $\mathbb{P}^n$ is the set of equivalence classes

$\{[x_0,\ldots,x_n],\ x_i \in k$ with at least one $x_i \neq 0$, $(x_0,\ldots,x_n) \sim (y_0,\ldots,y_n)$ iff $\exists \lambda \in k^*$ such that $x_i = \lambda y_i\}$.

Let $\mathbb{P}^n_i = \{[x_0,\ldots,x_n]$ with $x_i \neq 0\}$. It can be seen that the map $\phi_i : \mathbb{P}^n_i \longrightarrow k^n$ given by $\phi_i([x_0,\ldots,x_n]) = (\frac{x_0}{x_i},\ldots,\frac{\widehat{x_i}}{x_i},\ldots,\frac{x_n}{x_i})$ is well-defined and bijective. Thus, $\mathbb{P}^n_i$ can be given the structure of an affine variety. (In fact, of $\mathbb{A}^n$, the affine n-space). The algebra of functions is $k\left[\frac{X_0}{X_i},\ldots,\frac{\widehat{X_i}}{X_i},\ldots,\frac{X_n}{X_i}\right]$. It is easy to see that $\mathbb{P}^n = \bigcup_{i=0}^{n} \mathbb{P}^n_i$. Also, $\mathbb{P}^n_i \cap \mathbb{P}^n_j$ is a principal open set in $\mathbb{P}^n_i$ as well as in $\mathbb{P}^n_j$. Define $U \subseteq \mathbb{P}^n$ to be open iff $U \cap \mathbb{P}^n_i$ is open $\forall 0 \leq i \leq n$. It follows that $\mathbb{P}^n$ is a variety. This variety is called the projective space of dimension n.

One can define, for a vector space V of dimension $n+1$, $\mathbb{P}(V)$ in a similar manner, by choosing a basis. It is easy to check that the structure defined on $\mathbb{P}(V)$ is independent of the basis chosen.

A closed subvariety of a projective space is called a projective variety.

Our basic proposition is :

<u>Proposition 2</u>. A projective variety is complete.

<u>Proof</u>. In view of the proposition 1, it is enough to prove that a projective space $\mathbb{P}^n$ is complete. One proceeds by induction. $\mathbb{P}^o$, being a point, is complete. Now let W be any variety and $S \subseteq \mathbb{P}^n \times W$ be closed. We have to show that $p_2(S)$ is closed, where $p_2 : \mathbb{P}^n \times W \longrightarrow W$ is the projection map. It is clear that one may assume the following:
(1) W is affine, (2) S is irreducible. Let B be the function-algebra of the affine variety W. Let $\overline{B}$ be the function-algebra of the <u>irreducible</u> subvariety $\overline{p_2(S)}$ of W. ($\overline{B}$ is an integral domain). Consider $S \cap \mathbb{P}^n_i \times W = S_i$, $0 \leq i \leq n$.

If S_i is empty for some i, then $S \subseteq (\mathbb{P}^n - \mathbb{P}^n_i) \times W$. It is easy to see that $\mathbb{P}^n - \mathbb{P}^n_i$ can be canonically identified with $\mathbb{P}^{n-1}$. Hence by induction, it follows that $p_2(S)$ is closed. So let $S_i \neq \emptyset$ $\forall\, o \leq i \leq n$. Thus the function $\frac{X_i}{X_j}$ is not identically zero $\forall\, i,j$ (hence not identically ∞ $\forall\, i,j$). Consider the elements of $\bar{B}$ as functions on S_i in the obvious way and let $\overline{\frac{X_k}{X_j}}$ denote $\frac{X_k}{X_j}$ restricted to S_i. Since

$B\left[\frac{X_o}{X_i},\ldots,\frac{\hat{X_i}}{X_i},\ldots,\frac{X_n}{X_i}\right] = k\left[\frac{X_o}{X_i},\ldots,\frac{\hat{X_i}}{X_i},\ldots,\frac{X_n}{X_i}\right] \otimes B$ is the function algebra of $\mathbb{P}^n_i \times W$, it follows that $\bar{B}\left[\overline{\frac{X_o}{X_i}},\ldots,\overline{\frac{\hat{X_i}}{X_i}},\ldots,\overline{\frac{X_n}{X_i}}\right]$ is that of S_i. Also, S_i is irreducible. Hence $\bar{B}\left[\overline{\frac{X_o}{X_i}},\ldots,\overline{\frac{\hat{X_i}}{X_i}},\ldots,\overline{\frac{X_n}{X_i}}\right] = C_i$ is an integral domain.

Claim: The quotient field F_i of C_i is independent of i. This is easy to see, since the function $\overline{\frac{X_k}{X_j}} \in C_j = \overline{\frac{X_k}{X_i}} \Big/ \overline{\frac{X_j}{X_i}} \in F_i$. Thus we have the following:

$$\bar{B} \xrightarrow[p_2^*]{} \bar{B}\left[\overline{\frac{X_o}{X_i}},\ldots,\overline{\frac{\hat{X_i}}{X_i}},\ldots,\overline{\frac{X_n}{X_i}}\right] \hookrightarrow F \; (= F_i).$$

Let $q \in \overline{p_2(S)}$. Hence $e_q : \bar{B} \longrightarrow k$ is a k-algebra homomorphism. By the lemma in proposition 1 of 1.13, it follows that e_q can be extended to a k-algebra homomorphism $\emptyset : R \longrightarrow k$, where $R \subseteq F$ is a valuation ring of F (as obtained in the proof of the lemma).

Claim: $R \supseteq C_{i_o}$ for some i_o. Let i_o be an index $(0 \leq i_o \leq n)$ such that the set $J_{i_o} = \left\{ \begin{matrix} 0 \leq j \leq n \\ j \neq i_o \end{matrix} \overline{\frac{X_j}{X_{i_o}}} \in R \right\}$ is of maximal cardinality. If $j \notin J_{i_o}$; i.e. $\overline{\frac{X_j}{X_{i_o}}} \notin R$ then $\frac{X_{i_o}}{X_j} \in R$. ($R$ is a valuation ring). Also, $k \in J_{i_o} \Rightarrow k \in J_j$. Thus $i_o \in J_j - J_{i_o}$ which contradicts the fact that J_{i_o} has the maximal cardinal. This proves that $J_{i_o} = \{0,1\ldots n\}$. Hence $\underline{R \supseteq C_{i_o}}$. Hence we get $\emptyset : C_{i_o} \longrightarrow k$. Hence $\exists\, p \in S_{i_o} = S \cap \mathbb{P}^n_{i_o} \times W$ such that $\emptyset = e_p$. It follows from the compatibility of e_p and e_q that $p_2(p) = q$ or $q \in p_2(S)$. This proves the fact that $p_2(S)$ is closed. Hence $\mathbb{P}^n$ is complete.

We next prove:

Proposition 3. Products of projective varieties are again projective varieties.

Proof. Clearly, it is enough to prove that the product of two projective spaces is a projective variety. So consider two projective spaces $\mathbb{P}^n$ and $\mathbb{P}^m$. Consider the map

$$\phi : \mathbb{P}^n \times \mathbb{P}^m \longrightarrow \mathbb{P}^{(n+1)(m+1)-1} \quad \text{given by :}$$

$$\phi([x_o, \ldots, x_n], [y_o, \ldots, y_m]) = [x_o y_o, \ldots, x_o y_m, x_1 y_o, \ldots, x_n y_m] .$$

It can be easily seen that ϕ is a morphism, in fact, an isomorphism onto the image. Hence the proposition follows.

We now generalise the concept of projective spaces: We define:

Grassmannian Varieties. Let V be an n+1 -dimensional vector space. Let $G_d(V)$ be the set of all d-dimensional subspaces of $V (0 \leq d \leq n+1)$. Then $G_d(V)$ is called a Grassmannian-variety. Consider the map $\phi : G_d(V) \longrightarrow \mathbb{P}(\wedge^d V)$ given by: $\phi(W) = (v_1 \wedge v_2 \wedge \ldots \wedge v_d)$, where $\{v_1, \ldots, v_d\}$ is a basis for the d-dimensional subspace W of V. It can easily be checked that ϕ is well-defined.

Proposition 4. If $G_d(V)$, ϕ are as above, then ϕ is injective and $\phi(G_d(V))$ is closed in $\mathbb{P}(\wedge^d V)$. Thus $G_d(V)$ can be given the structure of a projective variety (which is complete by proposition 2). Its dimension is $d(n+1 - d)$.

Proof. The proof may be found in Borel's book (page 239).

We define another type of projective varieties viz.

Flag-varieties. Let V be an $(n+1)$-dimensional space over k. Let $\mathcal{F}(V)$ be the set of all flags of V. (To recall, a flag is a sequence $0 = V_0 \subsetneq V_1 \subsetneq \ldots \subsetneq V_n = V$ of subspaces of V such that $\dim V_i = i \ \forall 1 \leq i \leq n$).

Prcposition 5. If V, $\mathcal{F}(V)$ are as above, the natural map $\eta : \mathcal{F}(V) \longrightarrow G_0(V) \times \ldots \times G_n(V)$ is injective. Also, $\eta(\mathcal{F}(V))$ is closed. Thus $\mathcal{F}(V)$ can be given the structure of a (complete) projective variety.

The proof is easy but it will be omitted (See Borel's book, page 241).

2.11. Quotients. Let G be an affine algebraic group and $H \subsetneq G$ be closed subgroup.

Definition. A pair (π, V), where V is a variety and $\pi : G \longrightarrow V$ is a morphism, is called a quotient for $G/_H$ if the following conditions hold:

(1) The fibres of π are just the cosets of H in G. (hence π is surjective).

(2) π is open.

(3) If $U \subsetneq V$ is open, then $\pi^*(k[U]) = k[\pi^{-1}(U)]^H$, the algebra of all functions on $\pi^{-1}(U)$, constant on the cosets of H, i.e. the fibres of π.

Example. Consider two affine algebraic groups U and V. Then (p_2, V) is a quotient for $U \times V/_U$ ($p_2 : U \times V \longrightarrow V$ is the projection).

It can be seen that if a quotient (π, V) for G/H exists, then it is unique. The morphism $\pi : G \longrightarrow V$ is universal among all the morphisms from G which are constant on cosets of H.

Orbit maps. Let G be an affine algebraic group acting transitively on a variety V. Let $v \in V$, $H = G_v = \{g \in G / g.v = v\}$ be the stabilizer of v. Then there exists a natural map $\pi : G \longrightarrow V$ given by $\pi(g) = g.v$. Clearly, π is a morphism and (1) above holds. The openness of π, (2) above, will be proved below (see appendix). Then (π, V) is a quotient for G/G_v if the condition (3) holds. When does this condition hold? We have:

Proposition 1. Let G, V, v, H, π be as above. Then (π, V) is a quotient for G/H, i.e. (3) holds, iff the differential map $(d\pi)_1 : T(G)_1 \longrightarrow T(V)_v$ is surjective. This map is always surjective if $p(k) = 1$ (Thus (π, V) is a quotient for G/H if $p(k) = 1$).

The proof of this proposition may be found in Borel's book (page 180).

A short description of tangent spaces and differentials is as follows:

Let V be an affine variety and A be its algebra of functions. Introduce the dual numbers $k[\epsilon]$ with $\epsilon^2 = 0$. Consider a k-algebra homomorphism $\alpha : A \longrightarrow k[\epsilon]$. This is of the form $\alpha = \beta + r.\epsilon$ where $\beta : A \longrightarrow k$ is a k-algebra homomorphism and $r : A \longrightarrow k$ is a linear map satisfying: $r(a.b) = \beta(a).r(b) + \beta(b).r(a)$ $\forall a, b \in A$. Now, $\beta = e_v$ for some $v \in V$. Then $r(a.b) = a(v).r(b) + b(v).r(a)$, so that r satisfies the rules for differentiation at the point v. The set of such r's with a fixed $\beta = e_v$ (i.e. $\{ r / e_v + r.\epsilon$ is a k-algebra homomorphism$\}$) is a vector space over k and is called the tangent space to V at v. It is denoted by $T(V)_v$. The set of all α's i.e.

the union of all $T(V)_v$'s is just the tangent bundle. If $f : V \longrightarrow W$ is a morphism of affine varieties, then f gives rise to a k-linear map $(df)_v : T(V)_v \longrightarrow T(W)_{f(v)}$ in a natural way: Let $r \in T(V)_v$, then $(e_v + r.\epsilon) \circ f^* = e_{f(v)} + (r \circ f^*).\epsilon$ is again a k-algebra homomorphism, hence $r \circ f^* \in T(W)_{f(v)}$. Define $(df)_v(r) = r \circ f^*$. Then df is called the differential of f.

In the case of an arbitrary variety one can define the tangent space at a point by using an arbitrary open affine neighbourhood of that point.

Remark. The condition " $d\pi$ is surjective" in Proposition 1 is often stated "π is separable". It rules out unwanted inseparability, which, e.g., prevents the map $\mathbb{A}^1 \longrightarrow \mathbb{A}^1$, $x \rightsquigarrow x^p$ (p = char $k \neq 0$) from being an isomorphism even though it is bijective.

Theorem 1. If G is an affine algebraic group and $H \subseteq G$ is a closed subgroup, then $G/_H$ exists as a quasi-projective variety (i.e. open subset of a projective variety). It can be realized via an orbit map in the projective space $\mathbb{P}(V)$, coming from a linear action of G on V (for some V). If H is also normal, then $G/_H$ is affine and becomes an affine algebraic group.

The proof consists of constructing an orbit map for which the conditions of proposition 1 hold. (See page 181 of Borel's book).

After these preparations, we come to a basic tool in the study of affine groups.

Borel's fixed point theorem. A connected, solvable affine group G acting on a non-empty complete variety V always has a fixed point.

Proof. We prove the theorem by induction on dim G. If dim G = 0, G = $\{e\}$

and then there is nothing to prove. So let $\dim G > 0$. Since G is solvable and $\neq \{e\}$, $DG \underset{\neq}{\subset} G$. Now DG is closed and hence is of smaller dimension. Thus, by the induction hypothesis, DG has a fixed point in V. Let W be the set of fixed points of DG ($W \neq \emptyset$). Clearly, W is invariant under the action of G. (DG is normal in G). Thus, one can assume (by taking W instead of V) that DG acts as identity on the whole of V. We now have a homogeneous situation (taking a <u>closed</u> orbit of G, which exists). Let $v \in V$, G_v = stabilizer of v in G. Then $G \xrightarrow[\eta]{} V$; $\eta(x) = x.v$ is a morphism of G which is constant on the cosets of G_v. Hence, by the universal property of $G \to G/G_v$, there exists a morphism $\phi : G/G_v \longrightarrow V$ given by $\phi(\bar{g}) = g.v$. ϕ is clearly bijective. Also, $G_v \supseteq DG$ and hence G_v is normal in G. Thus, G/G_v is an affine algebraic group. Now the theorem follows from the following proposition.

<u>Proposition 2</u>. Let $f : V_1 \longrightarrow V_2$ be a G-morphism of homogeneous G-varieties (G is a connected, affine algebraic group and acts transitively on V_1 and V_2) with finite fibres. If V_2 is complete then so is V_1.

Applying the proposition to the morphism $\phi : G/G_v \longrightarrow V$, we get that G/G_v is complete (since V is complete). But then G/G_v is a connected, affine group. Hence by proposition 1 of 2.10, $G/G_v = \{e\}$ or $G = G_v$. Thus G fixes v.

The proof of the proposition 2 is contained in the appendix.

As a corollary to the above theorem, we have:

<u>(Lie-Kolchin) Theorem</u>. (as stated in 2.8). Let G be a connected, solvable, (closed) subgroup of $GL(V)$. Then G acts on $\mathcal{F}(V)$ in a natural way. By

proposition 5 above, $\mathcal{Y}(V)$ is complete. Hence the theorem follows immediately from the fixed point theorem.

Appendix to 2.11. This appendix brings several complements to the development so far.

Proposition 1. Let G be an affine algebraic group; V_1, V_2 homogeneous spaces for G; and $\pi: V_1 \longrightarrow V_2$ a G-morphism (i.e. $\pi(g.v_1) = g.\pi(v_1) \forall g \in G$, $v_1 \in V_1$).

(a) π is open.

(b) If π has finite fibres and V_2 is complete, then V_1 is also complete.

Part (b) is the missing step (viz. proposition 2) of the proof of the fixed point theorem, while part (a) shows, as mentioned above, that orbit maps are always open, so that in particular, surjective morphisms of algebraic groups are always open.

We shall give the proof, assuming G to be connected, in several steps.

(1) We define a morphism (of affine varieties) $f: U_1 \longrightarrow U_2$ to be finite if $f(U_1)$ is dense in U_2 and $k[U_1]$ is integral over $f^*(k[U_2])$. The fact that every homomorphism : $f^*(k[U_2]) \longrightarrow k$ extends to $k[U_1]$ translates geometrically to : $f(U_1)$ is closed, and similarly for any closed subset of U_1; i.e. f is a closed map. Taking complements, we see that $f(U)$ is open for any open subset U (of U_1) which is made up of complete fibres.

(2) Let $f: V_1 \longrightarrow V_2$ be a morphism of irreducible varieties with $f(V_1)$ dense in V_2. Then for suitable open affine subsets V_1^1, V_2^1 of V_1, V_2 respectively,

we have: $f(V_1^1) \subseteq V_2^1$ and $f^1 : f/_{V_1^1}$ can be factored: $V_1^1 \xrightarrow{g} \mathbb{A}^r \times V_2^1 \xrightarrow{p_2} V_2^1$ with g finite and p_2 as usual. Clearly, we may assume that V_1 and V_2 are affine. Identify $B = k[V_2]$ with a part of $A = k[V_1]$ via f^*. Working in the quotient field $Q(A)$ of A, we see by Noether's lemma (page 4, Mumford's book) that there exist elements $x_1, \ldots, x_r \in Q(B)[A]$, indeed in A, algebraically independent over $Q(B)$, such that $Q(B)[A]$ is integral over $Q(B)[x_1, \ldots, x_r]$. In the equations expressing the integrality over $Q(B)[x_1, \ldots, x_r]$ for a finite generating set for $A/_B$, the coefficients are polynomials in $x_1, \ldots, x_r$, with coefficients in $Q(B)$, finite in number. Let $b \in B$ be a common denominator for all of these latter coefficients. Then $A\left[\frac{1}{b}\right]$ is integral over $B\left[\frac{1}{b}\right][x_1, \ldots, x_r]$ which is pure over $B\left[\frac{1}{b}\right]$. Geometrically, this means that we have our result with $V_1^1 = (V_1)_b$; $V_2^1 = (V_2)_b$.

(3) We observe that f^1 in (2) is open on sets made up of fibres since g is by (1) and p_2 certainly is.

(4) To prove (a), we may assume that $V_1 = G$, considered as a G-space in the obvious way; viz. $g(g') = g.g'$.

For: Let $v_1 \in V_1$, then

$$\begin{array}{ccc} V_1 & \xrightarrow{f} & V_2 \\ & {\scriptstyle \phi} \nwarrow \quad \nearrow {\scriptstyle \eta} & \\ & G & \end{array} \qquad \begin{array}{l} \phi(g) = g.v_1 \\ \eta(g) = g.f(v_1) \end{array}$$

is a commutative diagramme. Hence f is open if ϕ and η are open. ϕ and η are both G-morphisms.

Let S be an open set in $V_1 = G$. We must show that $f(S)$ is open. If H is the stabilizer of $v = f(1)$ in G, then $SH = \bigcup_{h \in H} Sh$ is also open, consists of complete fibres, and $f(SH) = f(S)$. Thus we may assume that S is made up of complete fibres. Now choose V_1^1, V_2^1 as in (2), and $\{x_i\}$ finite in G, such that $\bigcup_i x_i V_1^1 = V_1$. Then by (3), $f(S \cap x_i V^1)$ is open $\forall$ i, whence $f(S)$ is open, as required.

(5) Assume now as in (b). It is enough to show that $V_1 \times W \longrightarrow V_2 \times W$ is closed for any W, since V_2 is complete (for any W affine in fact). Choose V_1^1, V_2^1 as in (2) and $\{x_i\}$ as in (4). Since the fibres are finite, dim V_1 = dim V_2 and r = 0 above, so that $f^1 : V_1^1 \longrightarrow V_2^1$ is finite. Hence $V_1^1 \times W \longrightarrow V_2^1 \times W$ is also finite (if A is integral over B, then $A \otimes C$ is integral over $B \otimes C$), so that it is closed. The same holds for every $x_i V_1^1 \times W \longrightarrow x_i V_2^1 \times W$ so that $V_1 \times W \longrightarrow V_2 \times W$ is closed as required.

<u>Proposition 2</u>. Let $f : U \longrightarrow V$ be a dominant morphism of irreducible varieties. Then $\dim f^{-1}(v) = \dim U - \dim V$ for a dense open set of v's in V.

This is a weak version of Lemma 2 near the end of 1.13 that is enough for most of our purposes, e.g., for the proof of Proposition 2(c) of 1.13. For the present proof we may assume by step (2) above that f has the factorization $U \xrightarrow[g]{} \mathbb{A}^r \times V \xrightarrow[p_2]{} V$ with g a finite morphism. Now p_2^{-1} clearly raises dimensions of closed sets by exactly r and g^{-1} preserves them since algebraic extensions do not change transcendence degrees. Applying this to v, a point, and to V, we get $\dim f^{-1}(v) = r = \dim U - \dim V$.

<u>Proposition 3</u>. Let V be an irreducible variety. Then $\dim T(V)_v$ is finite, $\geq \dim V$, with equality for a dense open set of v's in V. (Such v's are called simple or nonsingular.)

This result, needed later, also uses Noether's theorem, refined as follows:

(*) If A is a finitely generated integral domain over a perfect field k then there exists a generating set $\{x_1, x_2, \ldots, x_n\}$ such that $\{x_1, x_2, \ldots, x_d\}$ (for some d) is algebraically independent over k and for each $i > d$, x_i is separable algebraic over $k(x_1,\dots,x_{i-1})$ with (monic) minimal polynomial, say $F_i(x_1,\dots,x_i)$, with coefficients in $k[x_1,\dots,x_{i-1}]$. The usual proof of Noether's theorem, as given, e.g., in Mumford's book, takes care of this refinement as well. To prove Proposition 3, we may assume V to be affine. We apply (*) with $A = k[V]_0$ (and $d = \dim V$). Let $v : x_i \longrightarrow v_i$ be a point of V. It readily follows (see 2.11) that $t : x_i \longrightarrow t_i$ is a tangent vector at v iff $\sum_j \left(\frac{\partial F_i}{\partial x_j}\right)_v t_j = 0$ for all $i > d$. Since this is a homogeneous system of $n-d$ equations for n unknowns, we get $n \geq \dim T(V)_v \geq d = \dim V$ with equality on the right on the set of points where some $(n-d)^{th}$ order minor of $\left(\frac{\partial F_i}{\partial x_j}\right)$ is nonzero, an open subset of V. As may be checked, the minor farthest to the right works out to $\prod_{i>d}\left(\frac{\partial F_i}{\partial x_i}\right)$ (it is lower triangular) which is not identically 0 because of the separability in (*) so that the above open set is nonempty, hence dense, as asserted.

2.12. Borel subgroups. Throughout this section, G will denote a connected, affine algebraic group.

Definition. A maximal connected solvable subgroup of G is called a Borel subgroup.

Remarks. (1) Borel subgroups exist for dimension reasons.

(2) A Borel subgroup is always closed, since the closure of a connected solvable subgroup is again a connected solvable subgroup of G.

Example. Let $G = GL(V)$ for some vector space V. Let B be the set of all upper triangular elements of $GL(V)$. Then B is a Borel subgroup of G.

The basic result is as follows:

Theorem 1. The Borel subgroups of G are conjugate to each other. If B is one of them, then G/B is complete.

Proof. Let B be a Borel subgroup of maximum dimension. From the theorem 1 of 2.11, we have: a representation $\rho: G \longrightarrow GL(V)$ and a point $W_1 \in \mathbb{P}(V)$ whose stabilizer is B and such that the resulting orbit map $G/B \longrightarrow G.W_1$, $xB \rightsquigarrow x.W_1$ is an isomorphism. Note that $\operatorname{Ker} \rho \subseteq B$, hence is also solvable.

Using Borel's fixed point theorem repeatedly, one has a flag $w : 0 = W_0 \subseteq W_1 \subseteq \dots W_n = V \in \mathcal{F}(V)$ such that $\rho(g)(W_i) \subseteq W_i$ $\forall 0 \leq i \leq n$, $\forall g \in B$. It is now clear that such a g necessarily belongs to B (i.e. $B = \{g \in G / \rho(g)(W_i) \subseteq W_i \ \forall 0 \leq i \leq n\}$). In other words, one has a map $G \xrightarrow[\phi]{} \hat{\mathcal{F}}(V)$, $\phi(g) = g.w$ such that $G_w = B$ and hence a map $G/B \longrightarrow Gw$. Since this map dominates the earlier one (the map: each flag goes to its vertex : $w \rightsquigarrow W_1$,) and vice-versa since the earlier one was an isomorphism, we see that this map is also an isomorphism. Let θ be the orbit Gw.

Claim: For any other orbit θ' (for the action of G on $\mathcal{F}(V)$), $\dim \theta' \geq \dim \theta$. Let $w' \in \theta'$. Then $G_{w'}$ fixes the flag w'. Hence $\rho(G_{w'})$ is solvable (being in upper triangular form). Since $\ker \rho$ is solvable, it follows that $G_{w'}$ is solvable and hence so is $G^o_{w'}$. Thus $G^o_{w'}$ is a connected solvable subgroup of G. Hence by maximality of $\dim B$, $\dim G^o_{w'} \leq \dim B$. Also, $\dim \theta' = \dim G - \dim G^o_{w'}$ (all the fibres of the morphism $G \longrightarrow \theta'$, $g \rightsquigarrow g.w'$ are of the same dimension) and similarly, $\dim \theta = \dim G - \dim B$. Hence $\dim \theta' \geq \dim \theta$, proving the claim. This proves that θ is of minimum dimension among the orbits. Hence by corollary 3 to proposition 1 of 1.13, θ is closed. Thus θ

is a closed, complete (in fact, projective) variety. But then $G/_B \xrightarrow{\sim} \theta$. This proves that $G/_B$ is complete projective <u>whenever</u> B has maximal dimension among the dimensions of the Borel subgroups.

Now, let B' be any other Borel subgroup. Let B' act on the complete variety $G/_B$ in a natural way; (i.e. $b'.(gB) = b'gB, b' \in B', g \in G$). Then by Borel's fixed point theorem, $\exists\, xB \in G/_B$ such that $b'xB = xB \;\forall b' \in B'$. This shows that $x^{-1}B'x \subseteq B$. But then B' is a Borel subgroup and hence, so is $x^{-1}B'x$. Hence $x^{-1}B'x = B$. This proves that <u>all Borel subgroups are conjugate</u> and in particular, have the same dimension. This proves that $G/_B$ is complete projective for <u>all</u> Borel subgroups.

<u>Corollary 1</u>. If P is a closed subgroup of G, then the following conditions are equivalent:

(a) $G/_P$ is complete.

(b) P contains a Borel subgroup.

<u>Proof</u>. (a) $\Longrightarrow$ (b). Let B be a Borel subgroup. Let B act on $G/_P$ by left translations. Then by B.F.P.T., (abbreviation for Borel's fixed point theorem) B has a fixed point xP. This clearly gives (b) because P then contains the Borel subgroup $x^{-1}Bx$.

(b) $\Longrightarrow$ (a). Let $P \supseteq B$, a Borel subgroup. Then the map $\eta: G \longrightarrow G/_P$ is constant on cosets of B and hence gives rise to a (surjective) morphism: $\bar{\eta}: G/_B \longrightarrow G/_P$. This shows that $G/_P$ is complete, because $G/_B$ is.

Thus: Borel subgroups are the 'smallest' among the set of closed subgroups P with $G/_P$ complete, i.e. among the "parabolic subgroups".

Corollary 2. The maximal tori of G are all conjugate to each other and so are the maximal, connected unipotent subgroups.

Proof. Let T, T' be two maximal tori. Since T is connected solvable, $T \subseteq B$, a Borel subgroup. Similarly, $T' \subseteq B'$. But then B and B' are conjugate to each other. Hence we may assume that T and T' are contained in some Borel subgroup B. It now follows from the theorem 2 of 2.8, that T and T' are conjugate in B. Again, if U is maximal connected unipotent subgroup, then $U \subseteq B$ for some Borel subgroup B (B is nilpotent by Kolchin's Theorem in 2.5). Now, by theorem 2 of 2.8, $U \subseteq B_u$. This shows that $U = B_u$. Now the conjugacy of such U's follows from the conjugacy of Borel subgroups.

Corollary 3. (a) If α is an automorphism (endomorphism) of G, identity on B, then α is the identity of G.

(b) $Z_G(B) \subseteq Z_G(G)$.

Proof. (a) Consider the morphism $\phi : G \longrightarrow G$ given by $\phi(x) = \alpha(x).x^{-1}$. Then ϕ is constant on the cosets of B. Hence by the universal property of the quotient morphism, ϕ factors to $\overline{\phi} : G/_B \longrightarrow G$ given by $\overline{\phi}(gB) = \phi(g) = \alpha(g).g^{-1}$. Now $G/_B$ is complete and irreducible while G is affine. Hence by proposition 1 of 2.10, $\overline{\phi}$ is constant and clearly this constant = e, the identity element. This proves that α = Identity on G.

(b) This follows by applying (a) to the inner automorphisms by elements of $Z_G(B)$.

Remark. In the same way, one can prove that if G acts on an affine variety, then any point fixed by B is fixed by G.

Corollary 4. If B, a Borel subgroup, is nilpotent, then so is G. In fact, we prove G = B.

Proof. We proceed by induction on dim B. If dim B = 0, then $B = \{e\}$ and $G/B \xrightarrow{\sim} G$. Now G is affine, connected and G/B is complete. Hence $G = \{e\} = B$. So let $\dim B \geq 1$. Since B is nilpotent, there exists a closed subgroup $C \subsetneq Z_B(B)$ such that $\dim C \geq 1$. By corollary 3 above, $C \subsetneq Z_G(G)$. Hence G/C is an affine (connected) algebrai group. (Theorem 1 of 2.11).

Claim. B/C, which is a subgroup of G/C, is in fact a Borel subgroup.

Consider: $G \longrightarrow G/C \longrightarrow G/C\big/B/C$, the canonical morphism η. Then η is constant on cosets of B and hence factors through G/B. Thus, we have $G/B \xrightarrow[\bar{\eta}]{} G/C\big/B/C$; $\bar{\eta}$ is surjective. Thus $G/C\big/B/C$ is complete since G/B is so. B/C is already connected and solvable. Hence by corollary 1, it follows that B/C is a Borel subgroup of G/C. Now $\dim B/C < \dim B$ and B/C is again nilpotent. Hence by induction hypothesis, $G/C = B/C$. This shows G = B and completes the proof of the corollary.

2.13. Density and Closure.

Definition. A subgroup C of G is called a Cartan subgroup if $C = Z(T)^0$ for some maximal torus T in G. (We later prove that Z(T) is connected, so that C = Z(T)).

All Cartan subgroups are conjugate since all maximal tori are.

Proposition 1. If T, C are as above, then T is the unique maximal torus of C.

Proof. Clearly, T is maximal in C. Also, T is normal in C and by corollary 2 of 2.12, all maximal tori in C are conjugate. Hence the proposition follows.

Corollary 1. All Cartan subgroups are nilpotent.

Proof. Let $C = Z(T)^o$, T a maximal torus in G. Choose a Borel subgroup B of C such that $C \supseteq B \supseteq T$. Now T is a maximal torus in B. Hence by theorem 2 of 2.8, $B = T.B_u$. Also, T is normal in B. Hence the above is a direct product decomposition. It follows that B is nilpotent (since T and B_u are so). Hence by corollary 4 of 2.12, $C = B$ and C is nilpotent.

We now prove an important lemma.

Lemma 1 (Density). The union of the Cartan subgroups of G contains a dense open subset of G.

Proof. Fix a Cartan subgroup $C = Z(T)^o$, T a maximal torus. By the conjugacy of Cartan subgroups, one has to prove that $K = \bigcup_{g \in G} g C g^{-1}$ contains a dense open subset of G.

Let $S_o = \{(x, y) \in G \times G / x^{-1}yx \in C\}$. Clearly S_o is closed. S_o is also irreducible, being the image of $G \times C$ under the map $\theta : G \times G \longrightarrow G \times G$, $\theta(x, y) = (x, xyx^{-1})$. Again, S_o is made up of cosets of $C \times \{1\}$ in $G \times G$, since $(x, y) \in S_o$ implies that $(xc, y) \in S_o$ $\forall c \in C$. By using a theorem on compatibility of quotients and products (Borel's book, page 179), one has

$G \times G/_C \times \{1\} \xrightarrow{\sim} G/_C \times G$. Hence $\eta : G \times G \longrightarrow G/_C \times G$, given by $\eta\,(g, g') = (gC, g')$, is just a quotient map. η is open and hence the image of a closed subset consisting of complete cosets of $C \times \{1\}$ is closed. Thus $S = \eta\,(S_o)$ is closed. It is irreducible as well since S_o is. $\left[S = \left\{(xC, y)/x^{-1}yx \in C\right\}\right]$.

Now consider $p_1 : G/_C \times G \longrightarrow G/_C$ and $p_2 : G/_C \times G \longrightarrow G$. Clearly, $p_1(S) = G/_C$ and the fibre of xC is xCx^{-1}. It now follows that the fibres of p_1 have the same dimension.

Hence, dim C = dimension of a fibre

$$= \dim S - \dim p_1(S)$$
$$= \dim S - \dim G/_C\,.$$

Hence <u>dim S = dim C + dim G/C = dim G.</u> Again, consider $p_2 : S \longrightarrow \overline{p_2(S)}$. S is an e'pais in S, hence $p_2(S)$ is an e'pais, i.e. $p_2(S)$ contains an open subset of $\overline{p_2(S)}$. Note that $p_2(S) = K(= \bigcup_{g \in G} gCg^{-1})$. Hence the lemma is proved if we prove $\overline{p_2(S)} = G$.

<u>Claim:</u> $\dim p_2(S) = \dim G$. This clearly shows $\dim \overline{p_2(S)} = \dim G$ and hence $\overline{p_2(S)} = G$, both being irreducible and of the same dimension.

We observe that C has the following property:

(*) There exists $t \in C$ such that $\left\{xC/x^{-1}tx \in C\right\}$ is finite.

<u>For</u>: By the corollary to the proposition in 2.7 there exists $t \in T \subseteq C$, such that $Z_G(t) = Z_G(T)$. If C' is any Cartan subgroup containing t, then $t \in C' = Z_G(T')^o \Longrightarrow T' \subseteq Z_G(t) = Z_G(T)$. Hence $T' \subseteq Z_G(T)^o = C \Longrightarrow T = T'$ by proposition 1 above. Hence $C' = C$.

<u>Thus:</u> C is the unique Cartan subgroup containing t. Now, $x^{-1}tx \in C \Longrightarrow t \in xCx^{-1} \Longrightarrow xCx^{-1} = C \Longrightarrow x \in N_G(C) \Longrightarrow x \in N_G(T)$, since T is the unique maximal torus of C. Also, $N_G(T)^o = Z_G(T)^o = C$ (by corollary 2

of the theorem of 2.7). It now follows that the number of distinct cosets xC with $x^{-1}tx \in C \leq$ order of $N(T)/N(T)^o$ which is finite. Hence C satisfies the property (*).

Coming back to the claim, we see that for the morphism $p_2/_S : S \longrightarrow p_2(S)$ ($p_2 : G/_C \times G \longrightarrow G$; $S = \{(xC, y)/x^{-1}yx \in C\}$), the fibre over $t \in C$ is finite. Hence $\dim S = \dim p_2(S) = \dim G$. This proves the claim and hence the lemma.

(Note that we have used Lemma 2 of 1.13 which gives: (**) **For a dominant morphism** $f : U \longrightarrow V$, $\dim U = \dim V$ it some fibre is finite, non-empty. But we have not proved this lemma in these notes. We have, however, proved Proposition 3, Appendix to 2.11, which yields (**) with some replaced by most. This is enough for our proof here since from Proposition (b) of 2.7, it readily follows that most fibres of $p_2/_S$ above are finite. A similar remark applies to our later applications of Lemma 2 of 1.13.)

Remark. Let D be any subgroup of G satisfying the property (*). Then from the proof of the above lemma, it clearly follows that $\bigcup_{g \in G} gDg^{-1}$ contains a dense open subset of G.

Lemma 2. (Closure). Let G act on a variety V. Let H be a closed subgroup of G and $U \subseteq V$ be a closed subset of V, invariant under the action of H. Assume $G/_H$ to be complete. Then $G.U$ is closed.

Proof. Let $S = \{(xH, v)/x^{-1}v \in U\}$, $S \subseteq G/_H \times V$. Since $h(U) \subseteq U \ \forall \ h \in H$, it follows that S is well defined. Since U is closed, S is closed in $G/_H \times V$. Hence $p_2(S)$ is closed in V, since $G/_H$ is complete. But then $p_2(S) = G.U$, which proves this lemma.

<u>Theorem 1.</u> (a) The union of the Borel subgroups of G is all of G i.e. every element x of G is contained in a Borel subgroup (i.e. in a <u>connected solvable</u> subgroup).

(b) Every semisimple element is contained in a torus. Every unipotent element is contained in a <u>connected</u> unipotent group.

<u>Proof.</u> (a) Since every Cartan subgroup is contained in a Borel subgroup, it follows from the Density lemma that $\bigcup_{\text{all Borel } B'} B'$ contains a dense open subset of G. Now let G act on itself by conjugation. Choose a Borel subgroup B. Take V = G, H = B, U = B in the closure lemma. Then $G.B = \bigcup_{g \in G} gBg^{-1} = \bigcup_{\text{all Borel } B'} B'$ is closed. This clearly proves (a).

(b) Since every element is contained in a Borel subgroup, which is connected and solvable, (b) immediately follows from theorem 2 of 2.8.

<u>Remark.</u> For an arbitrary algebraic group G (connected), it is not true that each of its elements is contained in a <u>connected abelian</u> subgroup.

Consider G to be the group of upper triangular 2×2 matrices of determinant one (k of char. 0). Let $x = \begin{bmatrix} -1 & 1 \\ 0 & -1 \end{bmatrix}$. Then it can be easily checked that the only connected subgroup containing x is G itself and G is <u>not abelian</u>.

We state here an extension of the above theorem.

<u>Theorem 2.</u> Every surjective endomorphism of an affine group keeps some Borel subgroup invariant. In particular, every automorphism does so.

For the proof of this and related matters, see A.M.S. Memoir No. 80. (In particular, for an inner automorphism $i_x, i_x(B) = B$ for some B. Hence $x \in B$

(since every Borel subgroup is its own normalizer)).

Corollary 0. If B is a Borel subgroup of G, a connected group, then $Z(B) = Z(G)$.

Proof. We have $Z(B) \subseteq Z(G)$ by Theorem 1, Corollary 3(b). If $x \in Z(G)$, then $x \in B'$, some Borel, by Theorem 2, hence $x \in B$ since B' is conjugate to B.

Corollary 1. (Further closure). Let G act on an affine variety V. Let $v \in V$ be such that $G_v \supseteq$ a maximal torus. Then $G.v$, the orbit of v, is closed.

Proof. Let $G_v \supseteq T$, a maximal torus. Let $T \subseteq B$, a Borel subgroup. Then $B = B_u.T$ (by theorem 2 of 2.8); $S = B.v = B_u.T.v = B_u.v$, since $T.v = v$. Thus S is closed, being the orbit under the action of a unipotent group (by the proposition of 2.5). Hence by the closure lemma, as S is invariant under B, $G.v = G.S$ is closed.

Corollary 2. Let G be an affine group. (a) Any semisimple conjugacy class, or more generally (b) any conjugacy class meeting a Cartan subgroup is closed.

Proof. By theorem 1 above, every semisimple element is contained is a torus and hence in a Cartan subgroup. Thus (b) clearly implies (a).
Now to prove (b), let $x \in C$, a Cartan subgroup. Let $C = Z_G(T)^{o}$, T a maximal torus. Let G act on G by conjugation. Then $G_x = Z_G(x) \supseteq T$. Hence by corollary 1 above, the orbit of x, i.e. the conjugacy class, is closed.

Corollary 3. If S is a torus in G, then $Z_G(S)$ is connected.

Proof. Let $x \in Z_G(S)$ be arbitrary. Fix a Borel subgroup B. Since $x \in B'$, a Borel subgroup, and B' acting on $G/_B$ has a fixed point (by B.F.P.T.), it follows that x has a fixed point in $G/_B$. Let W be the set of all fixed points of x . Then W is a non-empty closed subset of $G/_B$. Thus W is complete also. Again, $x \in Z_G(S)$ and hence S keeps W invariant. Hence S has a fixed point in W (by B.F.P.T.). In other words, $\exists\ gB \in G/_B$ such that $S.gB = gB$ and $x.gB = gB$. Hence both S and x belong to the same Borel subgroup $gBg^{-1} = B'$, say. Now $x \in Z_{B'}(S)$ which is connected by theorem 2 of 2.8. Thus $Z_{B'}(S) \subseteq Z_G(S)^o$ and hence $x \in Z_G(S)^o$. This shows that $Z_G(S) = Z_G(S)^o$. This proves the corollary.

Remarks. (1) The first part of the argument shows that any connected solvable group and any element in its centralizer can be put in a Borel subgroup.

(2) A Cartan subgroup $C = Z_G(T)^o$ is, in fact, $= Z_G(T)$.

(3) The above corollary is not true in case S is a commutative group consisting of semisimple elements. Consider $G = PSL_2(\mathbb{C})$, $x = \begin{bmatrix} i & 0 \\ 0 & -i \end{bmatrix}$. It can be seen that $Z_G(x)$ = set of diagonal matrices $\bigcup$ set of matrices of the form: $\begin{bmatrix} 0 & a \\ -a^{-1} & 0 \end{bmatrix}$ with $a \in \mathbb{C}^*$. Thus $Z_G(x)$ is not connected.

Corollary 4. Let $t \in G$ be semisimple. Then $Z_G(t)/Z_G(t)^o$ consists of semisimple elements, i.e. every unipotent element in $Z_G(t)$ is in $Z_G(t)^o$.

Proof. Let $u \in Z_G(t)$ be an unipotent element. Let $x = t.u$. Clearly, this is the Jordan decomposition of x. Choose a Borel subgroup B containing x. It follows that $t, u \in B$ also. (Let $x = t'.u'$ be the Jordan decomposition in B. $t', u' \in B$. Then by proposition 3 of 2.4, $x = t'.u'$ is the decomposition in G as well. Hence by uniqueness, $t = t'$, $u = u'$). Now $u \in Z_B(t)$, which is connected by theorem 1 of 2.8. Hence $Z_B(t) \subseteq Z_G(t)^o$, so that $u \in Z_G(t)^o$. This proves the corollary.

2.14 Bruhat Lemma. (a) Let G be a connected algebraic group. Let T be a maximal torus in G. Let B be a Borel subgroup of G such that $B \supseteq Z(T) \supseteq T$. Then the canonical map $i : {}_{Z(T)}\backslash N(T) / {}_{Z(T)} \longrightarrow {}_{B}\backslash G / {}_{B}$ is a bijection. $\left(i\,(Z(T)\,.\,n.\ Z(T)) = B.n.B,\ n \in N(T) \right)$.

This lemma is sometimes expressed in a different form, viz. (a') Any two Borel subgroups of a connected algebraic group have a maximal torus in common.

It can be seen that (a) $\Longleftrightarrow$ (a').

For: (a) $\Longrightarrow$ (a'). Let B_1, B_2 be two Borel subgroups. Then $\exists\, g \in G$ such that $gB_1g^{-1} = B_2$. Choose a maximal torus T such that $T \subseteq Z(T) \subseteq B_1$. Then by (a), $\exists\, n \in N(T)$ and $b_1, b_1' \in B_1$ such that $g = b_1.n.b_1'$. Now, $b_1.T.b_1^{-1} \subseteq B_1$. Also, $b_1Tb_1^{-1} = (gb_1'^{-1}.n^{-1}).T\,(n\,b_1'.g^{-1}) \subseteq gB_1g^{-1} = B_2$. This proves (a'). Conversely, (a') $\Longrightarrow$ (a). Pick $T \subseteq C \subseteq B$, $x \in G$. Then by (a'), ${}^{x}B = xBx^{-1}$ and B contains a maximal torus T'. Now T, T' are conjugate in B. Hence ${}^{bx}B \cap B \supseteq T$ for some $b \in B$. Hence ${}^{(bx)^{-1}}T, T$ are conjugate in B. Therefore ${}^{b'.(bx)^{-1}}T = T$ for some $b' \in B$. Thus $b'(bx)^{-1} \in N = N(T)$. This shows: $x \in B.N.B = \bigcup_{w \in W} Bn_wB$. For uniqueness, let $Bn_wB = Bn_{w'}B$. Then $bn_w = n_{w'}b'$ with $b, b' \in B$. Assume $t \in T$, arbitrary. Hence ${}^{bn_w}t = {}^{n_{w'}b'}t$. The left side equals ${}^{n_w}t.v$ $(v \in B_u)$ since $B = B_u.T$ (semidirect) and T is Abelian. The right side is thus in B and equals ${}^{n_{w'}}t.{}^{n_{w'}}v'$ $(v' \in B_u)$. Thus ${}^{n_{w'}}v' \in B_u$. Hence ${}^{n_w}t = {}^{n_{w'}}t$, and $w = w'$ as required. It would be nice if someone could supply at this stage of the development a simple proof of (a) or (a'), and also for the miraculous fact that W acting on X is always a group generated by reflections. We shall have to use these results without proof.

We prove the lemma for the classical groups.

(i) $G = SL_n$.

Let B be the group of upper triangular matrices. Let T be the group of diagonal matrices. Then the following can easily be verified: $Z(T) = T$; $N(T)$ is the group of monomial matrices (i.e. each row and column contains exactly one non-zero element). $N(T)/_{Z(T)}$ is isomorphic to the permutation group S_n. We now prove that any element $g \in G$ can be written as $b.w.b'$ with $b, b' \in B$ and $w \in N(T)$. Choose $b \in B$ such that the total number of zeros appearing at the beginning of the rows of $b.g$ is maximal. It can be easily seen that in this case, the numbers of zeros at the beginning of the various rows of $b.g$ are all distinct (since otherwise we could add a multiple of one row to a higher one, i.e. multiply on the left by some element of B, and increase the number of zeros) and hence are $0, 1, \ldots, n-1$ (in some order). Again by suitably multiplying on the left by an element w of N(T), the above order can be made $0, 1, \ldots, n-1$. It is clear that $w.b.g$ is in B. This proves that $g = b_1 w_1 . b_1'$ for some $b_1, b_1' \in B$; $w_1 \in N(T)$. Thus the map $i : Z(T)\backslash^{N(T)}/_{Z(T)} \longrightarrow B\backslash^{G}/_{B}$ is surjective. Next, let $w, w' \in N(T)$ such that $b.w.b' = w'$ for $b, b' \in B$. i.e. $b' = w^{-1}.b^{-1}.w'$. For a matrix x, define Supp $x = \left\{(i,j)/x_{ij} \neq 0\right\}$; $x = (x_{ij})$. In the above case, it can be easily checked that Supp $b' =$ Supp $(w^{-1}b^{-1}w') \supseteq$ Supp $(w^{-1}w')$ (since b^{-1} is super diagonal). But then b' is super diagonal and $w^{-1}w'$ is monomial. It immediately follows that $w^{-1}w'$ is diagonal i.e. $\in T$. This proves the injectivity of i. This completely proves the lemma for SL_n.

(ii) $G = Sp_{2n}$

Let J denote the $n \times n$ matrix $\begin{bmatrix} 0 & . & . & 1 \\ 0 & . & 1 & 0 \\ . & . & . & . \\ 1 & 0 & . & 0 \end{bmatrix}_n$. Let $M = \begin{bmatrix} 0 & J \\ -J & 0 \end{bmatrix}$. Define $Sp_{2n} = \left\{A \in SL_{2n} \middle/ AMA^t = M\right\}$. In other words, Sp_{2n} is the set of fixed points in SL_{2n} of the automorphism $\sigma : SL_{2n} \longrightarrow SL_{2n}$, given by $\sigma(A) = M.(A^t)^{-1}.$ M^{-1}.

Now, in SL_{2n}, the Bruhat lemma can be refined to: Every element of Bn_wB is uniquely expressible in the form $u.n_w.b$ with $b \in B$ and $u \in U_w = U \cap n_w . Un_w^{-1}$ (U, U^- respectively the unipotent upper triangular, lower triangular groups). Now σ keeps B, N, T (of SL_{2n}) invariant. e.g. let $x \in B$. Then $\sigma(x) = M.(x^{-1})^t.M^{-1}$. Now $x^{-1} \in B$ again, so let $x = \begin{bmatrix} X & Y \\ 0 & Z \end{bmatrix}$. Then

$\sigma(x) = \begin{bmatrix} J\,Z^t.J & -J.Y^t.J \\ 0 & J\,X^t\,J \end{bmatrix} \in B$ again. Similarly for N and T. Thus if $x \in Sp_{2n}$, $x = u.n_w.b$ in SL_{2n} .. (*), then $x = \sigma(x) = \sigma(u).\sigma(n_w).\sigma(b)$. Now from uniqueness of *, we get : $\sigma(n_w) \in Tn_w$, $\sigma(u) = u$, $\sigma(b) = b$ (mod T). Thus the Bruhat decomposition in SL_{2n} leads to one in Sp_{2n}. If we label the basis elements of the space on which Sp_{2n} is acting by indices $n, n-1, \ldots 1, -1, \ldots, -n$, then the Weyl group of Sp_{2n} works out to be the group of those permutations π (on $2n$ symbols) such that $\pi(-i) = -\pi(i)\ \forall i$, i.e. the octahedral group.

(iii) If we replace $-J$ by J in M above, we obtain a proof for SO_{2n}, and with a slight modification, one for SO_{2n+1} as well. Hence the Bruhat lemma is proved for the classical groups.

Remarks. SL_n acts on $\mathbb{P}^{n-1}$ and $\mathcal{F}(V)$ in a natural way. (V is an n-dimensional vector space). A simplex σ is an ordered sequence $\{p_1, \ldots, p_n\}$ of linearly independent points in $\mathbb{P}^{n-1}$. A flag $0 = W_0 \subsetneq W_1 \subsetneq \ldots \subsetneq W_n = V$ and a simplex $\{p_1 \ldots p_n\}$ are said to be incident if there exists a permutation π of n symbols such that W_i is generated by $\{p_{\pi(1)}, \ldots, p_{\pi(i)}\}\ \forall\ 1 \leq i \leq n$. Now the Bruhat decomposition in form (a') for SL_n can be stated geometrically as follows: Any two flags are incident with some simplex. In fact, this is the

form in which the Bruhat lemma was originally proved. The reader may wish to construct a purely geometric proof of this statement.

Chapter III

Reductive and semisimple algebraic groups, regular and subregular elements

Let G be a connected algebraic group. Let R denote the maximal, connected, solvable, normal subgroup of G. (R exists for dimension reasons and is clearly unique). R is called the radical of G.

3.1. Definitions. (1) G is said to be reductive if $R_u = \{1\}$ (i.e. R does not contain a non-trivial unipotent element), or, equivalently, if R is a torus.

(2) G is said to be semisimple if $R = \{1\}$, or, equivalently, if G does not contain an abelian, normal subgroup of positive dimension.

Remarks. (1) If G is reductive, then the radical R, which is a torus, is in the centre of G. (Since R is normal, $N_G(R) = G$ (= $N_G(R)^0$). By corollary 2 to theorem of 2.7, $Z_G(R)^0 = N_G(R)^0 = G$. Hence R is central.)

(2) For any (connected) group G, G/R is semisimple (this is clear) and G has a decomposition $G = R.G_1$, where G_1 is semisimple. (Levi decomposition).

Examples. (1) GL_n is a reductive group. The radical R of it has eigen-spaces V_i on the underlying space V. Since R is normal, G permutes these V_i's and hence keeps their sum V_0 invariant. Since G acts irreducibly on V, it follows that $V_0 = V$. Thus R is diagonalizable, hence a torus, and G is reductive. In fact, since G is connected, G must fix the individual V_i's, hence there is only one such V_i. Thus R consists of scalar matrices. It is easy to see that $GL_n = R.SL_n$ is the decomposition of GL_n as in remark (2) above.

(2) SL_n, Sp_{2n}, SO_n are semisimple. (A proof similar to that above works.)

In the next 3 sections, G will denote a connected semisimple algebraic group.

3.2. Main Theorem on semisimple groups.

Let T be a maximal torus in G.

Definition. A character α of T is called a root of G (with respect to T) if there exists a morphism of algebraic groups $x_\alpha : G_a(k) \longrightarrow G$ ($G_a(k)$ is the additive group of k) such that the following conditions are satisfied:

(1) x_α is an isomorphism onto the image X_α, which is an unipotent group, normalised by T.

(2) $t.x_\alpha(r).t^{-1} = x_\alpha(\alpha(t)r) \; \forall t \in T, \; r \in k$.

Henceforth, we shall write X(T) additively (till 3.4).

Let R be the set of roots of G. Let $V = X(T) \otimes_{\mathbb{Z}} \mathbb{R}$. Then V is a vector space over $\mathbb{R}$ of dimension = dim T. Identify X(T) with $X(T) \otimes 1$. Let $W = N(T)/_T$. For a semisimple group G, it can be proved that $Z(T) = T$ for a maximal torus T. Hence W is a finite group and is called the Weyl group of G relative to T. Clearly, W acts on X(T) and hence on V as a group of automorphisms. Choose a positive definite inner product on V, invariant under W. (Such an inner product exists, since W is finite). Denote it by

(,). Then the following results are true:

Theorem. $R \subsetneq V$ forms a root-system which is reduced. X_α and $X_{-\alpha}$ generate a subgroup which is isomorphic to SL_2 or PSL_2. This subgroup contains an element w_α of $N(T)$ which acts on V as the reflection relative to α (i.e. $w_\alpha . x = x - \frac{2(\alpha, x)}{(\alpha, \alpha)} . \alpha$). The Weyl group W is generated by $\{w_\alpha, \alpha \in R\}$. Choose a basis for the root system R. Let U be the group generated by $\{X_\alpha, \alpha > 0, \in R\}$. Let U^- be the group generated by $\{X_\alpha, \alpha < 0, \in R\}$. Let $B = T.U$, $B^- = T.U^-$ (Note that T normalises U, U^-.) Then U is a maximal, connected, unipotent subgroup of G, B is a Borel subgroup and the canonical morphism: $\prod_{\alpha > 0} X_\alpha \longrightarrow U$ is an isomorphism of varieties. (The product is taken in any order.) The canonical morphisms: $T \times U \longrightarrow B$ and $U^- \times B \longrightarrow U^-. B$ are also isomorphisms of varieties. $U^-. B$ is a dense open subset of G and is called the Big Cell. (Clearly, analogous statements are true in case of U^-).

We omit the proofs (which are quite long and may be found in § 13-14 of Borel's book), but shall use these results in what follows.

As an example, the reader may wish to verify these facts for the group $G = SL_n$. Take T as the diagonal group. For each i, j, $i \neq j$, $\alpha(i,j) : T \longrightarrow k$ given by $\alpha(i,j)(\text{diag}(t_1, \ldots, t_n)) = t_i . t_j^{-1}$ is a root. ($x_{\alpha(i,j)}(r) = I + r. E_{ij}$ is the corresponding morphism.) A root $\alpha(i,j)$ is positive if $j > i$. Here, $U^-. B$ consists of the products of subdiagonal, unipotent elements and upper diagonal elements. Any such product has the $i \times i$ minor in the upper left-hand corner not equal to zero, $(1 \leq i \leq n)$. Conversely, any such matrix may be uniquely so factored, the factors being polynomials in the co-ordinates and the reciprocals of these minors. (These facts may be verified by induction on n.)

A short description of root-systems is as follows: Let V be a finite dimensional vector-space. A symmetry s_α with respect to a vector $\alpha \neq 0$ in V is an automorphism of V which keeps a hyperplane pointwise fixed and takes α to $-\alpha$. A root system R in V is a finite set of non-zero generators along with a symmetry s_α for each $\alpha \in R$ such that

(1) $s_\alpha(R) \subseteq R \quad \forall \alpha \in R$,

(2) For each $\alpha, \beta \in R$, $\beta - s_\alpha(\beta)$ is an integral multiple of α.

A root system R is said to be reduced if $\alpha, t\alpha \in R \Rightarrow t = \pm 1$. The group W generated by $\{s_\alpha, \alpha \in R\}$ is a finite group and is called the Weyl group of R. If (,) is a positive definite inner product on V, invariant under W, then the symmetry s_α is given by : $s_\alpha(x) = x - \dfrac{2(\alpha, x)}{(\alpha, \alpha)} \cdot \alpha$, $x \in V$.

A basis for a root system R is a subset S of R such that (1) S is a basis for the vector space V and (2) any $\alpha \in R = \sum_{\beta \in S} m_\beta \cdot \beta$, m_β are integers with the same sign. A basis always exists and for any two bases S, S' of R, $\exists\, w \in W$ such that $w(S) = S'$.

A complete description may be found in J-P. Serre's book: Algebra de Lie semisimples complexes (Chapitre V). A comprehensive treatment may be found in Bourbaki's book. See also the appendix of Lectures on Chevalley groups.

3.3 Some Representation Theory.

Let $G \xrightarrow{\rho} GL(V)$ be a representation of G (V finite dimensional). For a character $\lambda \in X(T)$, define $V_\lambda = \left\{ v \in V / \rho(t).v = \lambda(t).v \ \forall t \in T \right\}$. λ is called a weight of the representation ρ if $V_\lambda \neq \{0\}$. Since T is diagonalizable, $V = \sum_{\lambda \in X(T)} V_\lambda$ (in fact, a direct sum).

Proposition 1. Every character of T is a weight for some representation of G.

Proof. Given $\alpha \in X(T)$, choose $f \in k[G]$ such that $f/_T = \alpha$. Since T acts semisimply on $k[G]$, there exist $f_i (1 \leq i \leq r) \in k[G]$ and α_i, $(1 \leq i \leq r) \in X(T)$ such that $f = \sum_{i=1}^{r} f_i$ and $t^* f_i = \alpha_i(t).f_i \ \forall t \in T$, $1 \leq i \leq r$. Now

$\alpha(t) = f(t) = (t^* f)(1) = \sum_{i=1}^{r} (t^* f_i)(1) = \sum_{i=1}^{r} f_i(1).\alpha_i(t)$. Hence $\alpha = \sum_{i=1}^{r} f_i(1)\alpha_i$.

By the linear independence of distinct characters, $\alpha = \alpha_i$ for some i. This proves the proposition.

Proposition 2. Let $G \xrightarrow{\rho} GL(V)$ be a representation of G. Let λ be a weight of ρ and $v \in V_\lambda$. Let α be a root of G (as defined in 3.2 above). Then $x_\alpha(r).v = \sum_{i=0}^{\infty} r^i.v_i$, the sum being finite, $v_i \in V_{\lambda + i\alpha}$ independent of $r \in k$, and $v_o = v$.

Proof. Since $x_\alpha(r).v$ is a polynomial in r, with coefficients in V, we have:

$$x_\alpha(r).v = \sum_{i=0}^{\infty} r^i.v_i, \text{ a finite sum.} \quad (\forall r \in k).$$

We have to show that $v_i \in V_{\lambda + i\alpha}$ and $v_o = v$. Let $t \in T$. Then

$$t.(x_\alpha(r).v) = \sum_{i=0}^{\infty} r^i.tv_i. \text{ Also,}$$

$$\begin{aligned} t.(x_\alpha(r).v) &= (t.x_\alpha(r).t^{-1})(tv) \\ &= x_\alpha(\alpha(t).r)\lambda(t).v \text{ as } v \in V_\lambda \\ &= \sum_{i=0}^{\infty} \alpha(t)^i.r^i.\lambda(t).v_i . \end{aligned}$$

Hence $\lambda(t). \sum_{i=0}^{\infty} \alpha(t)^i.r^i.v_i = \sum_{i=0}^{\infty} r^i.(tv_i)$. $\forall r \in k$, $t \in T$. It follows that

$\lambda(t).\alpha(t)^i.v_i = t.v_i \ \forall\, t \in T$. i.e. $v_i \in V_{\lambda+i\alpha}$. Setting $r = 0$, we get $v = v_0$.

Lemma. Let $G \xrightarrow[\rho]{} GL(V)$ be a representation of G. Let $v \in V_\lambda$ and $\alpha \in R$ (i.e. α be a root of G.) Then $w_\alpha . v \in V_{w_\alpha(\lambda)}$.

Proof. $t.w_\alpha v = w_\alpha(w_\alpha^{-1} t w_\alpha).v = w_\alpha . \lambda(w_\alpha^{-1}.t.w_\alpha).v$

$= w_\alpha(\lambda)(t).\ w_\alpha v \ \forall\, t \in T.$

Hence by definition, $w_\alpha . v \in V_{w_\alpha(\lambda)}$.

As an immediate consequence of this lemma, we have: V_λ and $V_{w(\lambda)}$ have the same dimension for $w \in W$.

Proposition 3. Let $\lambda \in X(T)$ and $\alpha \in R$. Then $\frac{2(\lambda,\alpha)}{(\alpha,\alpha)}$ is an integer.

Proof. Choose a representation $\rho : G \longrightarrow GL(V)$ such that λ is a weight for ρ. (This exists by proposition 1). By proposition 2, $\sum_{i \in \mathbb{Z}} V_{\lambda+i\alpha}$ is invariant under X_α and $X_{-\alpha}$ and hence under w_α by the theorem of 3.2. (w_α belongs to the group generated by X_α and $X_{-\alpha}$.) In particular, $w_\alpha . v \in \sum_{i \in \mathbb{Z}} V_{\lambda+i\alpha}$ for $v \neq 0 \in V$ (Such v exists.) But $w_\alpha . v \in V_{w_\alpha(\lambda)}$ by above lemma. It follows that $w_\alpha(\lambda)$ is of the form $\lambda + i\alpha$ for some $i \in \mathbb{Z}$. Hence $w_\alpha(\lambda) = \lambda - \frac{2(\lambda,\alpha)}{(\alpha,\alpha)}.\alpha = \lambda + i\alpha$ (for some $i \in \mathbb{Z}$). This proves the result.

If we denote the lattice $\left\{\lambda \in V \,\middle|\, \frac{2(\lambda,\alpha)}{(\alpha,\alpha)} \in \mathbb{Z} \ \forall \alpha \in R\right\}$ by $L^*(R)$, then proposition 3 simply says: $X = X(T) \subseteq L^*(R)$. If $L(R)$ denotes the lattice generated by R i.e. the set $\left\{\sum_{\alpha \in R} n_\alpha . \alpha,\ n_\alpha \in \mathbb{Z}\right\}$ then we have: $\underline{L(R) \subseteq X \subseteq L^*(R)}$.

Definitions. (1) G is said to be simply connected if $L^*(R) = X$.

(2) G is said to be adjoint if $L(R) = X$.

Let $\{\alpha_1, \dots, \alpha_n\}$ be a basis for the root-system R. Define $\lambda_i \in V$ by
$\frac{2(\lambda_i, \alpha_j)}{(\alpha_j, \alpha_j)} = \delta_{ij}$ $1 \leq i \leq n$, $1 \leq j \leq n$. It can be easily checked that $\{\lambda_1, \dots, \lambda_n\}$ is a basis of V and $L^*(R)$ is just the lattice generated by it. $\lambda_1, \dots, \lambda_n$ are called the fundamental weights with respect to the basis $\{\alpha_1, \dots, \alpha_n\}$ of R. (To prove that $\lambda_i \in L^*(R)$, we have to use the fact that $\left\{\frac{2\alpha_j}{(\alpha_j, \alpha_j)}, 1 \leq j \leq n\right\}$ is a basis of the dual root-system R^* of R).

<u>Proposition 4</u>. If G, R, V, $\{\lambda_i\}_{1 \leq i \leq n}$ are as above, then the following conditions are equivalent:

(1) G is simply connected.

(2) $\lambda_i \in X$, $1 \leq i \leq n$

(3) There exist $\mu_1, \dots, \mu_n$ in X such that $w_{\alpha_j}(\mu_i) = \mu_i - \delta_{ji}\alpha_j$ $1 \leq i, j \leq n$.
Further $\mu_i = \lambda_i$ in (3).

The proof is clear.

<u>Remark</u>. The following condition is equivalent to the above 3 conditions: If $\pi : G' \longrightarrow G$ is an isogeny (i.e. having a finite kernel) with G' connected such that π is an isomorphism on connected unipotent subgroups, then π itself is an isomorphism.

<u>Proposition 5</u>. SL_n is simply connected.

<u>Proof</u>. We choose R, $\{\alpha_i\}$ as given in 3.2 above. (To recall, a positive root is of the form $\alpha(i,j)$ with $i < j$). Define $\mu_i \in X(T)$ by μ_i (diag $(t_1, \dots, t_n)$) = $t_1 \dots t_i$ $1 \leq i \leq n-1$. (μ_i is indeed a character.) It is easily checked that $w_{\alpha(j,j+1)} \cdot (\mu_i) = \mu_i - \delta_{ji}\alpha(j,j+1)$ $\forall j$. Hence by proposition 4, SL_n is simply connected.

Remarks. The groups Sp_{2n}, $Spin_n$ are also simply connected. The groups SO_{2n+1}, PSL_n are adjoint, while the groups SO_{2n} are neither. The groups G_2, F_4, E_8 are simultaneously simply connected and adjoint.

We now state the fundamental result on the classification of semisimple groups.

Theorem. Given an abstract root system R and any lattice X between L(R) and $L^*(R)$, there exists, up to isomorphism, a unique semisimple group over any algebraically closed field k such that R and X are realized as above (relative to any maximal torus.)

3.4. Representation Theory (continued).

Dominant Weights. Let G, X, R, V be as in 3.3. Choose a basis S to R. An element $\lambda \in V$ is said to be dominant if $(\lambda, \alpha^*) \geq 0$ for all positive roots α ($\alpha^* = \frac{2\alpha}{(\alpha,\alpha)}$). It follows that such λ's form a cone C with $\lambda_1, \ldots, \lambda_n$ (the fundamental weights) as a basis. (i.e. every element $\lambda \in C = \sum a_i \lambda_i$, $a_i \geq 0$).

Define an order $\geq$ on V by : $\lambda \geq \mu$ iff $\lambda - \mu$ is a sum of positive roots. It can be checked that $\geq$ is a partial order. Moreover, given $\lambda \in C \cap L^*(R)$, there exist only finitely many μ's, also in $C \cap L^*(R)$, such that $\lambda \geq \mu$. (For: If $\lambda \geq \mu$ then $(\lambda,\lambda) - (\mu,\mu) = (\lambda+\mu, \lambda-\mu) \geq 0$. Thus μ is confined to a bounded part of space as well as to a lattice.)

We now prove a basic result:

Fundamental Theorem of Representation Theory. Let G be a semisimple group, T a maximal torus and B the Borel subgroup corresponding to the positive roots (with respect to a basis). Then we have:

(a) If (π, V) is an irreducible representation of G, then there exists a unique line D which is fixed by B, the corresponding character λ is uniquely determined and is dominant, and all other weights of this representation are of strictly lower order (relative to the above partial order).

(b) Two irreducible representations are isomorphic iff the corresponding dominant characters, called highest weights, are equal.

(c) Given a dominant character λ on T, there exists an irreducible representation (which is unique up to isomorphism because of (b)) (π, V) such that the corresponding highest weight is λ.

Proof. (a) Let (π, V) be an irreducible representation. By B.F.P.T., B fixes a flag. Hence there exists a line D which is kept invariant by B. Choose $v \in D$, $v \neq 0$. Let λ be the corresponding character on T (i.e. $t.v = \lambda(t).v \ \forall t \in T$). Since V is irreducible, it follows that V is spanned by G.v. Since the 'big cell' $U^{-}B$ is dense in G, it follows that V is spanned by $U^{-}B.v$ i.e. by $U^{-}.v$ (as B acts as scalars on v). Consider an element $x_{-\alpha}(C_{\alpha})$ of U^{-}. By proposition 2, $x_{-\alpha}(C_{\alpha}).\ v = v + \sum_{i \geq 1} C_{\alpha}^{i}.v_i$ with $v_i \in V_{\lambda - i\alpha}$. It now follows that for any $u^{-} \in U^{-}$, $u^{-}.v = v + w$ where $w \in \bigoplus_{\mu < \lambda} V_{\mu}$. Thus,

$$V = k.v. \oplus (\bigoplus_{\mu<\lambda} V_{\mu}) \ (\text{each } \mu = \lambda - \sum_{\alpha > 0} n_{\alpha}.\alpha,\ n_{\alpha} \geq 0). \qquad (*)$$

Note that $V_{\lambda} = k.v = D$. If D' is any other line fixed by B, then arguing in the same way as before, we get a decomposition (*) of V, with some weight λ' instead of λ. It now follows that $\lambda' \geq \lambda$ and $\lambda \geq \lambda'$. Since V_{λ} is one dimensional, $V_{\lambda} = D = D'$. Thus D is the unique line kept invariant by B. Again, for any α simple, $w_{\alpha}(\lambda)$ is also a weight of π. But $w_{\alpha}\lambda = \lambda - (\lambda, \alpha^{*})\alpha$.

Hence $(\lambda, \alpha^*) \geqslant 0$ and λ is dominant.

(b) Let V_1, V_2 be two irreducible representations with the same highest weight λ. Let $V = V_1 \oplus V_2$. Choose non-zero vectors $v_i \in V_i$ $(i = 1, 2)$ corresponding to the dominant character λ. Let $v = v_1 + v_2 \in V$. Let W be the G-subspace generated by v. We have, $W = \langle G.v \rangle = \langle U^-.B.v \rangle = \langle U^-.v \rangle$ $= k.v +$ lower weight spaces, since $v = v_1 + v_2$ corresponds to the weight λ. $W \cap V_2$ is a G-submodule of the irreducible module V_2 and does not contain v_2 ($v_2 \notin W$ by above). It follows that $W \cap V_2 = \{0\}$. Hence $p_1 : W \longrightarrow V_1$ is injective. Since $p_1.v = v_1 \neq 0$ and V_1 is irreducible, p_1 is surjective also. Hence W is isomorphic to V_1. Similarly W is isomorphic to V_2. Hence V_1 and V_2 are isomorphic.

(c) The proof of this part may be found in the author's "Lectures on Chevalley group" (p. 210). It is so vital for our further development that we shall indicate a proof: In $A = k[G]$, let A_λ be the space of functions f which satisfy (*): $f(b^-x) = \lambda(b^-).f(x)$ for all $b^- \in B^-$, $x \in G$. ($\lambda \in X(T)$ can be extended to a character on $B^- = U^-.T$ as T normalizes U^-.) Suppose we know that A_λ is non-zero. G acts on A_λ via right translations, locally finitely. Let V_λ be an irreducible submodule. (V_λ is finite dimensional). By (a), there exists a highest weight vector f corresponding to some highest weight λ'. On the big cell $U^-B = U^-.TU$, we have : $f(U^-.tu) = \lambda(t).f(u)$ by (*). In particular, $f(U^-.t) = \lambda(t).f(1)$. Also, since f is a highest weight vector, $f(u^-.t) = \lambda(t).f(u^-) = \lambda(t).f(1)$. Now $f(1) \neq 0$, since otherwise $f = 0$. Hence $\lambda = \lambda'$ and V has λ as its highest weight. The proof that A_λ is not zero, or equivalently, that the function f defined on the big cell by $f(u^- tu) = \lambda(t)$ exists on G as a polynomial, requires further argument, which may be found in the above mentioned book. We see, incidently, that the irreducible representations of

G are all induced representations, induced from one dimensional representations of B.

This proves the theorem completely.

Henceforth, we write X(T) multiplicatively and reserve the addition sign for functions (in $k[G]$ or $k[T]$).

The Weyl group W defines an equivalence relation among the characters on $T(\lambda \sim w(\lambda),\ \lambda \in X(T), w \in W)$. It can be easily seen that each equivalence class contains exactly one dominant character. For the class $[\mu]$, define Symm$[\mu]$ to be the sum (as functions on T) of all characters belonging to it. For a representation $G \xrightarrow{\rho} GL(V)$, we define $X_\rho : G \longrightarrow k$ by $X_\rho(g) = \text{Trace}(\rho(g))$. Clearly $X_\rho \in k[G]$. Consider X_ρ on T. For each class $[\mu]$ of (equivalent) characters, V_μ has a constant dimension, $\mu \in [\mu]$. It follows that on T, X_ρ is just a sum of Symm $[\mu]'$s. Let ρ_λ be irreducible with λ as the highest weight. Since any weight of ρ_λ, other than λ itself, is of order strictly lower than λ and λ has multiplicity 1, it follows from the above that X_{ρ_λ} on T is given by

$$X_{\rho_\lambda} = \text{Symm}\,[\lambda] + \sum_{\substack{\mu < \lambda \\ \mu \text{ dominant}}} \text{Symm}\,[\mu] \quad \text{———} \ (*)$$

We denote X_{ρ_λ} by just X_λ and Symm $[\lambda]$ by just Symm λ. From (*), it immediately follows that

$$\text{Symm}\,\lambda = X_\lambda + \sum_{\substack{\mu < \lambda \\ \mu \text{ dominant}}} \pm X_\mu \quad \text{———} \ (**).$$

<u>As an example, we consider $G = SL_n$.</u> Let $\lambda_i \in X(T)$ be as defined in proposition 4 of 3.3. We can realize ρ_i in $\Lambda^i(k^n)$ with the usual action

$g.(v_1\wedge \dots \wedge v_i) = gv_1 \wedge \dots \wedge gv_i$. If $\{e_1, \dots, e_n\}$ is the standard basis of k^n, then $e_1 \wedge \dots \wedge e_i$ is a highest weight vector and the other standard basis vectors of $\wedge^i(k^n)$ are obtained by permutations. Hence, in this case, X_{λ_i} = Symm λ_i on T and hence $X_{\lambda_i}(g)$ is just the ith elementary symmetric polynomial in the eigenvalues of g. (This is clear if $g \in T$ and hence also if $g \in G$ arbitrary.)

Definition. $f \in k[G]$ is said to be a class-function if it is constant on the conjugacy classes. The set of all class functions is denoted by $C[G]$. e.g. Any character X_ρ, corresponding to a representation ρ, is a class-function (being a trace-function).

Theorem 2. Let G be a semisimple algebraic group, T, a maximal torus and W, the corresponding Weyl group.

(a) The restriction to T of the X_λ's (for dominant characters λ) form a linear basis of $k[T]^W$, whereas the X_λ's themselves form a linear basis of $C[G]$.

(b) If G is simply connected and $\{\lambda_i\}_{1\le i\le n}$ are the fundamental weights, then $\{X_{\lambda_i/T}\}_{1\le i\le n}$ (respectively $\{X_{\lambda_i}\}_{1\le i\le n}$) freely generate $k[T]^W$ (respectively $C[G]$) as k-algebra.

Proof. We first prove the statements concerning T and then deduce the corresponding results for G.

(a) Since Symm λ and X_λ are interrelated, by the equations (*), (**) above, it is sufficient to prove that $\{\text{Symm } \lambda\}$ is a basis of $k[T]^W$. Let $f \in k[T]^W$. Since T is diagonalizable, the characters of T span $k[T]$. So let

$f = \sum_{X \in X(T)} C_X . X$. Hence for $w \in W$, $f = wf = \sum_X C_X . wX = \sum_X C_{w^{-1}(X)} . X$. Now characters are linearly independent, hence $C_X = C_{w^{-1}(X)}$ $\forall X, \forall w$. This means that the elements of an equivalence class (of characters under action of W) occur with the same coefficient. Hence $f = \sum_X C_X$. Symm X and the Symms span $k[T]^W$. Further, let $\sum_X a_X$. Symm X = 0. Now, characters occuring in Symm X are distinct from those occuring in Symm $X'(X \not\sim X')$. Since characters are linearly independent, it follows that $a_X = 0 \; \forall X$. This proves that $\{\text{Symm } \lambda\}$ is a basis of $k[T]^W$. The relations (*) and (**) then imply that $\{X_\lambda\}$ is also so.

(b) If G is simply connected, the fundamental weights $\{\lambda_i\}$ are in fact characters. Let X_i denote X_{λ_i}. Now, for any λ dominant, $\lambda = \prod \lambda_i^{n_i}$ with $n_i \geq 0$. Since

$$X_\lambda = \text{Symm}\,\lambda + \sum_{\substack{\mu < \lambda \\ \mu \text{ dominant}}} \text{Symm}\,\mu = \lambda + \sum_{\substack{w.\lambda \neq \lambda \\ w \in W}} w.\lambda + \sum_{\substack{\mu < \lambda \\ \mu \text{ dominant}}} \text{Symm}\,\mu,$$

it can be easily seen that $X_\lambda - \prod X_i^{n_i}$ is a sum of X'_μs with $\mu < \lambda$. Since there exist only finitely many characters (dominant) which are less than λ, it follows, by repeated application of the above argument, that X_λ is a polynomial in X_i's. Again, let p be a polynomial in n variables with $p(X_1, \ldots, X_n) = 0$. We write p in the form $p = a.X_1^{r_1} \ldots . X_n^{r_n} + p_1$, where $X_1^{r_1} \ldots . X_n^{r_n}$ is the term of highest order. (We use a laxicographic order). Then by an argument similar to the one above, it follows that $a = 0$. Thus it can be proved that p is identically zero. Thus $X_1, \ldots, X_n$ generate freely $k[T]^W$ as k-algebra.

We now prove similar statements for $C[G]$. Consider the restriction map : $C[G] \xrightarrow{\phi} k[T]^W$, which is well defined. We claim that it is in fact, an isomorphism. Since X_λ, for λ a dominant character, is in $C[G]$, surjectivity is obvious from the above results for $k[T]^W$. Further, let $f \in C[G]$

such that $f/_T = 0$. Now for any x semisimple $\in G$, $gxg^{-1} \in T$ for some $g \in G$. Hence $f(x) = f(gxg^{-1}) = 0$ (f is a class-function). Thus f is zero on the set of semisimple elements, i.e. $f = 0$ on $\bigcup_{g \in G} gTg^{-1} = \bigcup_{C \text{ cartan}} C$ (since $T = Z(T)$) and $\bigcup_{C \text{ cartan}} C$ is dense in G (Density lemma of 2.13). It now follows that $f = 0$ on G. Hence ϕ is injective. Now the statements for $C[G]$ are obvious from those for $k[T]^W$.

This proves the theorem completely.

Note. Taking $G = SL_n$ and choosing co-ordinates properly on T = the group of diagonal matrices, we see that the above theorem is just the fundamental theorem on symmetric polynomials. In fact, the above method of proof is completely parallel to one of the standard proofs of that theorem.

This theorem has some interesting corollaries:

Corollary 1. If $f \in C[G]$ and $x \in G$, then $f(x) = f(x_s)$.

Proof. For any representation ρ, we may write $\rho(x)$ in superdiagonal form with $\rho(x_s)$ as its diagonal (by the lemma of 2.1). Thus $X_\rho(x) = X_\rho(x_s)$. Now the corollary (1) follows from (a) of the theorem.

Corollary 2. (a) The semisimple classes are in one-one correspondence with elements of $T/_W$ (i.e. set of orbits of T under the action of W.)

(b) If G is simply connected, then $T/_W$ is isomorphic to the affine-space $\mathbb{A}^r$ under the map:

$$\phi : T/_W \longrightarrow \mathbb{A}^r \; ; \; \phi(\bar{t}) = (X_1(t), \ldots, X_r(t)), \bar{t} \in T/_W .$$

<u>Proof.</u> (a) Consider the map $\eta : T/_W \longrightarrow$ (Conjugacy classes of semisimple elements) given by : $\eta(\bar{t}) = [t]$. Clearly, η is well defined and surjective. Let $t, t' \in T$ such that $[t] = [t']$. i.e. $\exists\, g \in G$ such that $gtg^{-1} = t'$. Now, gTg^{-1} and T are contained in $Z_G(t')^o$ and are maximal tori there. Hence $\exists\, h \in Z_G(t')$ such that $hgTg^{-1}h^{-1} = T$. Thus $hg \in N(T)$. Also, $hgtg^{-1}h^{-1} = ht'h^{-1} = t'$. In other words, $t \sim t'$ under the action of W. This proves the injectivity of η and hence (a).

(b) Consider the map $\phi : T \longrightarrow \mathbb{A}^r$, given by: $\phi(t) = (X_1(t), \ldots, X_r(t))$.

We prove: (1) $\phi^*(k[\mathbb{A}^r]) = k[T]^W$.

(2) Fibres of ϕ are just the orbits under W.

From the theorem above, (1) is clear.

To prove (2), we observe the following fact: If x, y are two elements of T which lie in <u>different</u> orbits, then there exists a function $f \in k[T]^W$ such that $f(x) = 0$, $f(y) \neq 0$. <u>For:</u> The orbits of x and y are finite, hence closed subsets of T. Hence the corresponding ideals I and J have no common zero and so their sum is $k[T]$. Write $1 = i+j$, $i \in I$, $j \in J$. Then i is 0 on the orbit of x and 1 on the orbit of y. Let $f = \prod_{w \in W} w.i$. Then f clearly satisfies the required properties. This proves that whenever $\phi(t) = \phi(t')$, t and t' belong to the same orbit. The converse is clearly true. Further, $k[T]$ is integral over $k[T]^W$. ($g \in k[T]$ satisfies $\prod_{w \in W} (X - w.g) = 0$, which is a monic polynomial in X with coefficients in $k[T]^W$). Thus a homomorphism of $k[T]^W$ into k can be lifted to a homomorphism of $k[T]$ into k. Using (1), it is now easy to see that ϕ is onto. Thus (b) is proved and hence so is the corollary.

Corollary 3. Let x, y be semisimple elements in G. Then the following statements are equivalent:

(1) x and y are conjugate.

(2) $X_\rho(x) = X_\rho(y)$ for every irreducible representation ρ.

(3) $\rho(x)$ is conjugate to $\rho(y)$ in $GL(V_\rho)$ for every irreducible representation ρ.

If G is simply connected, then (2) and (3) are replaced by :

(2') $X_i(x) = X_i(y) \ \forall\ 1 \leq i \leq r$.

(3') $\rho_{\lambda_i}(x)$ is conjugate to $\rho_{\lambda_i}(y)$ in $GL(V_{\rho_{\lambda_i}}) \ \forall\ 1 \leq i \leq r$.

Proof. (1) $\Longrightarrow$ (3) $\Longrightarrow$ (2) is clear.

(2) $\Longrightarrow$ (1). Clearly, one may assume that $x, y \in T$. Since $X_\rho(x) = X_\rho(y)$ for every irreducible representation ρ, the theorem shows that $f(x) = f(y) \forall f \in k[T]^W$. As seen in the proof of corollary (2), $k[T]^W$ separates orbits of W. It now follows that x and y belong to the same orbit of W i.e. x and y are conjugate.

In case of G being simply connected, (1) $\Longleftrightarrow$ (2') $\Longleftrightarrow$ (3') is proved in exactly the same way.

Remark. It is, however, not known whether a similar result holds for other elements of G.

Corollary 4. $x \in G$ is unipotent iff $X_\lambda(x) = X_\lambda(1) \ \forall$ dominant character λ i.e. the variety of unipotent elements is closed and is defined by equations $\{X_\lambda(x) = X_\lambda(1); \lambda \text{ a dominant character}\}$.

A similar result follows, in case of simply connected groups, with X'_λ s replaced by X'_{λ_i} s.

Proof. x is unipotent if $x_s = 1$ iff $X_\lambda(x_s) = X_\lambda(1)$ $\forall$ dominant characters λ iff $X_\lambda(x) = X_\lambda(1)$ $\forall$ dominant characters λ. (We use corollary (1) and (3) above).

Corollary 5. For a semisimple group G, a conjugacy class is closed iff it is semisimple.

Proof. $\Leftarrow$. This is already proved in corollary 2 to theorem 1 of 2.13. Another proof is given as follows: Fix x_0 (semisimple) in the conjugacy class. Consider a faithful representation of G. Let $S = \{x \in G | x$ satisfies the minimal polynomial of x_0 and $X_\lambda(x) = X_\lambda(x_0)$ for all dominant $\lambda\}$. S is closed and contains the conjugacy class of x_0. Conversely, if $x \in S$ then x is semisimple since its minimal polynomial has distinct roots, hence is conjugate to x_0 by corollary 3.

$\Rightarrow$. This implication follows from a general lemma:

Lemma. The closure of every conjugacy class of G contains, along with each element, its semisimple part. (G reductive).

Granting this fact, our result follows. For: a closed conjugacy class will contain the semisimple part of one of its elements and hence will be semisimple itself.

Proof of lemma. Let S be a conjugacy class, $x \in S$. We can assume, after conjugation, that $x \in B = T.U$ and $x_s \in T$, $x_u \in U$. Now x_u is of the form: $\prod_{\alpha>0} x_\alpha(C_\alpha)$, $C_\alpha \in k$. Choose $t \in T$ such that $\alpha(t) = C$, pregiven in k^*, for every simple root α. (Such t exists by proposition 7 of 2.6).

Now $txt^{-1} = t.x_s.x_u.t^{-1} = x_s.t\, x_u t^{-1} = x_s.\prod_{\alpha>0} x_\alpha(C^{\text{ht}\,\alpha} C_\alpha)$, where ht α = $\sum n_i$, $\alpha = \sum n_i \alpha_i$, $(\alpha_1, \ldots, \alpha_n)$, basis of R. Thus, $\prod_{\alpha>0} x_\alpha(C^{\text{ht}\,\alpha} C_\alpha) \in x_s^{-1}.\,\text{cl}\, S = \text{cl}\, x_s^{-1} S$. This is true for every $C \in k^*$. Take any $f \in k[G]$ such that $f(\text{cl}\, x_s^{-1} S) = 0$. The map $C \rightsquigarrow f(\prod_{\alpha>0} x_\alpha(C^{\text{ht}\,\alpha} C_\alpha))$ is a polynomial map, which is zero for every $C \in k^*$. This means that it is identically zero. In particular, $f(e) = 0$, also. It now follows that $e \in \text{cl}(x_s^{-1} S) = x_s^{-1}\, \text{cl}\, S$. Hence $x_s \in \text{cl}\, S$. This proves the lemma and hence the corollary.

3.5. Regular elements. In this section G will denote a connected, reductive algebraic group which may or may not be semisimple.

Let G be a (reductive) group. Let R be its radical. Then R is a torus which is central in G. We also have, $G = R.S$, where S is the semisimple group uniquely determined as the commutator subgroup. One can now talk of the root system of G (viz. that of S). A maximal torus of G is of the form $R.T'$ where T' is a maximal torus of S. It follows that $Z_G(RT') = R.T'$. Now, the results proved in the previous sections can be used here (with slight modifications, if required).

Definition. $x \in G$ is called a regular element if $Z_G(x)$ has the minimum dimension (among the centralizers of elements of G) or, equivalently, if $C(x)$, the conjugacy class of x, has the maximum dimension.

Clearly, regular elements exist for dimension reasons.

Proposition 1. The minimal dimension above is just r, the rank of G i.e. the dimension of a maximal torus.

Proof. Fix a maximal torus T. Choose $t \in T$ such that $Z_G(T) = Z_G(t)(= T)$. (Such an element exists by corollary to proposition of 2.7).

(1) Then $\dim T = r = \dim Z_G(t) \geq$ minimal dim.

(2) Next we prove: For any $x \in G$, $\dim Z_G(x) \geq r$. (This proves the proposition.) Let $x \in B$, a Borel subgroup. We have, $B = T.U$, T is a maximal torus and U is the unipotent part of B. Since $(B, B) \subseteq U$, it follows that $\dim B/[B, B] \geq \dim T = r$. Now, $\dim Z_G(x) \geq \dim Z_B(x)$ = codimension of $C_B(x)$ in $B \geq r$, since $C(x).x^{-1} \subseteq [B, B]$.

This proves the required result.

To prove (2), we give an alternate method: Define $S_2 = \{(x, y) \in G \times G / x, y$ lie in a common torus$\}$. Fix a torus T.

Then S_2 is the image of $G \times T \times T$ under the map $\phi : G \times T \times T \longrightarrow G \times G$, given by $\phi(x, t, t') = (xtx^{-1}, xt'x^{-1})$. Hence S_2 is irreducible. Let G_2 be the closure of S_2 in $G \times G$. Then G_2 is also irreducible. Consider the projection $p_1 : G \times G \longrightarrow G$. Since semisimple elements are dense in G and $(x, 1) \in S_2$ whenever x is semisimple, it follows that $p_1(G_2) = G$. For any function $f \in k[G]$, $f(x.y) = f(y.x)$ $\forall (x, y) \in S_2$ (since $x.y = y.x$). Now the function $\overline{f} \in k[G \times G]$ defined by $\overline{f}(x, y) = f(xy) - f(y.x)$ is zero on S_2 and hence on G_2. Since $f \in k[G]$ is arbitrary, it follows that $x.y = y.x$ $\forall (x, y) \in G_2$. In other words, for $x \in G$, the fibre of p_1 at x is contained in $Z_G(x)$. Further, choose t semisimple such that dim fibre at t = $\dim G_2 - \dim G$. (Such t exists). Now for any $x \in G$, $\dim Z_g(x) \geq$ dim fibre at $x \geq \dim G_2 - \dim G$ = dim fibre at

$t \geqslant \dim T = r$, since $T \subseteq$ fibre at t. This proves (2).

Remark. By the above method of proof, one can show: For any $x \in G$, $Z_G(x)$ contains an abelian subgroup of dimension $\geqslant r$. The proof runs as follows: Define S_i, $G_i (i = 3, 4 \ldots)$ as above. (e.g. define $S_3 = \{(x, y, z) \in G \times G \times G / x, y, z$ belong to the same torus$\}$). Then (1) the components of any element of S_i, hence also of G_i, commute with each other, and (2) the map $f_i : G_{i+1} \longrightarrow G_i$, $f_i(x_1, \ldots, x_{i+1}) = (x_1, \ldots, x_i)$ is surjective. (For: if $g_i : S_i \longrightarrow S_{i+1}, g_i(x_1, \ldots, x_i) = (x_1, \ldots, x_i, 1)$, then $f_i \circ g_i = 1$ on S_i, hence on G_i so that $G_i = f_i(g_i G_i) \subseteq f_i(G_{i+1})$). It follows that the map $p_1 \circ f_2 \circ \ldots \circ f_i : G_{i+1} \longrightarrow G$ is onto. Consider finite subsets $(y_1, \ldots, y_i)$ with $(x, y_1, \ldots, y_i) \in G_{i+1}$. Choose an i and such a subset such that $Z_G(y_1, \ldots, y_i)$ is minimal. (This is possible since G is noetherian). Let $z \in G$ such that $(x, y_1, \ldots, y_i, z) \in G_{i+2}$ i.e. let $z \in$ fibre of f_{i+1} at $(x, y_1, \ldots, y_i)$. By choice of y's, $Z_G(y_1, \ldots, y_i) = Z_G(y_1, \ldots, y_i, z)$. i.e. $z \in$ centre of $Z_G(y_1, \ldots, y_i)$. Also, $z \in Z_G(x)$. In other words, z ranges over an abelian subgroup of $Z_G(x)$. This subgroup, being a fibre of f_{i+1}, has dimension $\geqslant r$ by our earlier argument, whence our assertion.

As an immediate corollary, one gets: If x is regular, then $Z_G(x)^o$ is abelian. However it is not known whether the converse is true or not.

Proposition 2. For $G = SL_n$ or GL_n,

(a) A semisimple element is regular iff all of its eigenvalues are distinct from each other.

(b) A unipotent element is regular iff it is a 'single block' in the Jordan-Holder form.

(c) The following are equivalent:

(1) x is regular.

(2) The minimal polynomial of x is of degree n (i.e. the minimal polynomial = characteristic polynomial).

(3) $Z(x)$ is abelian.

(4) k^n is cyclic as $k[X]$ -module.

The proof of this proposition is straight forward and so is omitted.

We now characterize regular semisimple elements.

Proposition 3. For a semisimple $t \in G$, the following statements are equivalent:

(a) t is regular.

(b) $Z_G(t)^o$ is a torus, necessarily maximal.

(c) t belongs to a unique maximal torus.

(d) $Z_G(t)$ consists of semisimple elements.

(e) $\alpha(t) \neq 1$ for every root α relative to every, or to some, maximal torus containing t.

Proof. We choose a torus T and a Borel subgroup B such that $t \in T$, and $B = T.U$ is the decomposition as given in 3.2.

(a) $\Rightarrow$ (b). Since $T \subseteq Z_G(t)^o$ and $\dim T = r = \dim Z_G(t)^o$, it follows that $Z_G(t)^o = T$.

(b) $\Rightarrow$ (c). Let $t \in T'$, a maximal torus. Then $T \subseteq Z_G(t)^o$, which itself is a (maximal) torus. Hence $T' = Z_G(t)^o$. Thus t belongs to a unique maximal torus viz. $Z_G(t)^o$.

(c) $\Rightarrow$ (b). $T \subseteq Z_G(t)^o$. For any $g \in Z_G(t)^o$, $gtg^{-1} = t$, hence gTg^{-1} contains t. Hence by uniqueness, $gTg^{-1} = T$. Thus T is normal in $Z_G(t)^o$

which is connected. Hence T is central in $Z_G(T)^o$ so that $Z_G(t)^o \subseteq Z_G(T) = T$. This proves (b).

(b) $\Rightarrow$ (d). By corollary 4 to theorem 1 of 2.13, all the unipotent elements in $Z_G(t)$ belong to $Z_G(t)^o$. But then elements of $Z_G(t)^o$ are semisimple, $Z_G(t)^o$ being a torus. Hence $Z_G(t)$ does not contain any unipotent element. If $x \in Z_G(t)$, then $x_s, x_u \in Z_G(t)$ as well. Hence $x_u = 1$ and $x = x_s$ i.e. x is semisimple. This proves (d).

(d) $\Rightarrow$ (e). For a root system R of G with respect to T, let $\alpha(t) = 1$ for some $\alpha \in R$. Since $t.x_\alpha(c).t^{-1} = x_\alpha(\alpha(t).c) = x_\alpha(c)\ \forall c \in k$, it follows that $X_\alpha \subseteq Z_G(t)$. This clearly contradicts (d). Hence $\alpha(t) \neq 1$ for every root α.

(e) $\Rightarrow$ (b). We first prove: $Z_G(t)^o \cap U^-B \subseteq T$. Let $x \in Z_G(t)^o \cap U^-B$. Now x is of the form: $\prod_{\alpha>0} x_{-\alpha}(c_\alpha).t'.\prod_{\alpha>0} x_\alpha(d_\alpha)$, with $c_\alpha, d_\alpha \in k$, $t' \in T$.

Since, $x \in Z_G(t)$, $x = txt^{-1} = t(\prod_{\alpha>0} x_{-\alpha}(c_\alpha)).t'.(\prod_{\alpha>0} x_\alpha(d_\alpha)).t^{-1}$

$$= \prod_{\alpha>0} x_{-\alpha}(\alpha(t)^{-1}c_\alpha).t'.\prod_{\alpha>0} x_\alpha(\alpha(t).d_\alpha).$$

Hence by uniqueness, $\alpha(t)^{-1}.c_\alpha = c_\alpha$ and $\alpha(t).d_\alpha = d_\alpha$ $\forall \alpha>0$. Since $\alpha(t) \neq 1$ for any $\alpha \in R$, it follows that $c_\alpha = d_\alpha = 0\ \forall \alpha$. Thus $x = t' \in T$. Now U^-B is open in G (by theorem of 3.2). Hence $Z_G(t)^o \cap U^-B$ is open in $Z_G(t)^o$. Since $T \subseteq Z_G(t)^o \cap U^-B$ and $Z_G(t)^o$ irreducible, it follows that $Z_G(t)^o \cap U^-B = T$ is the whole of $Z_G(t)^o$. This proves (b).

(b) $\Rightarrow$ (a) is obvious.

This proves the proposition completely.

We now give a complete picture of $Z_G(t)$ for a semisimple element $t \in G$.

Proposition 4. Let $t \in T$, a maximal torus. Let R be the root system of G with respect to T. Then $Z_G(t)$ is the group G_1 generated by T, by those X_α's such that $\alpha(t) = 1$ and by those n_w's ($w \in W = N(T)/T$) such that $w(t) = t$ (i.e. $n_w t n_w^{-1} = t$). The identity component $Z_G(t)^o$ is generated by T and the X_α's (such that $\alpha(t) = 1$) and is reductive.

Proof. By Bruhat lemma, $G = \bigcup_{n_w \in N(T)} Bn_wB$, a disjoint union. Define $U_w = U \cap n_w.U^-.n_w^{-1}$. It is known that U_w is the group generated by X_α's such that $\alpha > 0$ and $w^{-1}\alpha < 0$. (Note that $n_w.X_\alpha n_w^{-1} = X_{w(\alpha)}$. We make use of a lemma which will be very useful later on.

Lemma. The map $\phi : U_w \times B \longrightarrow Bn_wB$ given by $\phi(u,b) = u.n_w.b$ is in fact an isomorphism of varieties. Hence, in particular, any element of Bn_wB is uniquely written as $u.n_w.b$ with $u \in U_w, b \in B$.

Proof of the lemma. Since $n_w.T.n_w^{-1} = T$, it follows that $Bn_w.B = U.n_w.B$. Further, $X_\alpha.n_w = n_w.X_{w^{-1}(\alpha)}$. Hence, whenever $\alpha > 0, w^{-1}(\alpha) > 0, X_\alpha.n_w \subseteq n_w.U$. It now follows that such X_α's can be 'passed' over n_w and absorbed in B. (Recall that $U = \prod_{\alpha>0} X_\alpha$ for every order of positive roots.) Thus $Bn_wB = U.n_w.B = U_w.n_w.B$. This proves the surjectivity of ϕ. For $u \in U_w, u = n_w.u^-.n_w^{-1}$ for some $u^- \in U^-$. Thus $u.n_w = n_w.u^-$. Now, let $\phi(u_1,b_1) = \phi(u_2,b_2)$, $u_i \in U_w, b_i \in B$, $i = 1,2$. i.e. $u_1n_w.b_1 = u_2.n_wb_2$. Hence $n_w.u_1^-.b_1 = n_w.u_2^- b_2$ for some $u_1^-, u_2^- \in U^-$. Thus $u_1^-.b_1 = u_2^-.b_2$. Since $U^- \times B \xrightarrow{\eta} U^-.B$, $\eta(u^-,b) = u^-.b$ is an isomorphism, it follows that $u_1^- = u_2^-$ and $b_1 = b_2$. Consequently $u_1 = u_2$. Hence ϕ is injective. It is easy to see that ϕ is in fact an isomorphism of varieties since the natural map: $U^- \times B \xrightarrow{\eta} U^-.B$ is. Thus, $G = \bigcup_{n_w} U_w.n_w.B$, a disjoint union.

Coming to the proof of the proposition, we observe that the group G_1 is contained in $Z_G(t)$. Now let $x \in Z_G(t)$ be arbitrary. Hence there exists n_w such that $x \in U_w.n_w.B$. Hence there exist unique $u \in U_w$, $b \in B$ such that $x = u.n_w.b$. Since $x \in Z_G(t)$, $x = txt^{-1} = tut^{-1}.tn_wt^{-1}.tbt^{-1}$. Now $tn_wt^{-1} = n_w.n_w^{-1}tn_wt^{-1} = n_w.t'.t^{-1}, t' = n_w^{-1}tn_w \in T$. Also, u is of the form: $\prod_{\substack{\alpha>0 \\ w^{-1}(\alpha)<0}} x_\alpha(c_\alpha)$.

Hence $tut^{-1} = \prod_{\substack{\alpha>0 \\ w^{-1}(\alpha)<0}} x_\alpha(\alpha(t).c_\alpha) \in U_w$ again.

Thus $x = (tut^{-1}).n_w.(t'.t^{-1}.tbt^{-1})$. Hence by uniqueness, $tut^{-1} = u, t'.bt^{-1} = b$. Thus $\alpha(t) = 1$ whenever $c_\alpha \neq 0$. Further, let $b = t''.u_o, u_o \in U$. Then $t''.u_o = t'.t''.u_o.t^{-1} = t'.t^{-1}.t''.t^{-1}(tu_ot^{-1})$. Again, by uniqueness, $t' = t$ and $u_o = tu_ot^{-1}$. Writing $u_o = \prod_{\alpha>0} x_\alpha(d_\alpha)$, one sees that $\alpha(t) = 1$ whenever $d_\alpha \neq 0$. Also, $t' = t$ gives $n_w.t.n_w^{-1} = t$. It is now clear that $x \in G_1$. Hence $Z_G(t) = G_1$.

Further, whenever $\alpha(t) = 1$, $(w^{-1}\alpha)(t) = \alpha(n_w.t.n_w^{-1}) = \alpha(t) = 1$. Hence any element of G_1 is of the form $g_2.n_w$, where $g_2 \in G_2$, the group generated by T and X_α's alone and n_w such that $n_w.t.n_w^{-1} = t$. Clearly, G_2 is a subgroup of finite index in G_1. Since G_2 is closed and connected, it immediately follows that $G_2 = G_1^o = Z_G(t)^o$.

For the reductivity, one may see 'Seminaire Chevalley, Vol. 2'. There it is also shown that the α's, for which $\alpha(t) = 1$, form a root system for G_2 and that T and the X_α's with $\alpha > 0$, $\alpha(t) = 1$ generate a Borel subgroup (of G_2). We observe that $Z_G(t)^o$ and G have the same rank since $T \subseteq Z_G(t)^o$.

<u>Remark.</u> t is regular iff $Z_G(t)^o = T$. This clearly shows the equivalence of (a) and (e) in the earlier proposition.

Corollary. Regular semisimple elements in a reductive group form an open set whose complement has codimension 1.

Proof. Choose a maximal torus T. Let R be the root-sustem relative to T. Let $f_o = \prod_{\alpha \in R} (\alpha - 1)$. Since $w \in W$ permutes R, it follows that $f_o \in k[T]^W$. Now by theorem 2 of 3.4, f_o extends to a unique class function $f \in C[G]$. We claim: $S = \{x / f(x) \neq 0\}$ is the set of regular semisimple elements. Clearly from (e) of proposition 3, S contains the set of all regular semisimple elements. Let $x \in S$. We can assume $x \in B$ (= T.U in usual notation). Again, by conjugating by a suitable element of B, one can assume $x = s.u, s \in T, u \in U$ is the Jordan decomposition of x. Since $f \in C[G]$, $f(x) = f(s)$. (See corollary 1 to theorem 2 of 3.4.) i.e. $\prod_{\alpha \in R} (\alpha - 1)(s) \neq 0$, i.e. $\alpha(s) \neq 1$ $\forall \alpha \in R$. Hence by proposition 3, s is regular. Hence $Z_G(s)$ consists of semisimple elements. Since $x \in Z_G(s)$, x is semisimple and hence regular (since u = 1). Thus S is an open dense set of G, whose complement has codimension 1 since it is defined by a single equation. (Check the last implication.)

Proposition 5. $x \in G$ is regular iff $x_u \in Z_G(x_s)^o$ is regular.

Proof. Let $Z_G(x_s)^o = G_1$. Then $Z_{G_1}(x_u) = Z_G(x) \cap G_1 \supseteq Z_G(x)^o$ (since $Z_G(x) \subseteq Z_G(x_s)$). Also, $Z_{G_1}(x_u)^o \subseteq Z_G(x)^o \cap G_1 = Z_G(x)^o$. Hence $Z_{G_1}(x_u)^o = Z_G(x)^o$. Further, whenever $x \in T$, a maximal torus, $T \subseteq Z_{G_1}(x_u)^o$ ($\subseteq G_1$). Hence T is a maximal torus in G_1 as well. The proposition now follows immediately.

3.6 Unipotent classes. To continue, we must know that regular unipotent elements exist. In case of characteristic 0, this was proved by Dynkin and Kostant, using Lie algebras. For arbitrary characteristic, this was proved by the author in Publi. Math. I.H.E.S. # 25 (1965) (§4). Using explicit calculation in the Lie algebra, Springer proved this result for 'almost all' characteristics. Here, we shall give an account of Richardson's method; he proved that under some conditions, a reductive group G contains only finitely many unipotent classes. (From this the existance of regular unipotent elements will follow).

We start with some preliminaries:

Lie algebra of an algebraic group G. We recall that the tangent space at a point $p \in G$ is defined to be the set of all k-algebra homomorphisms $e_p + \epsilon . \tau$ of $k[G]$ into the algebra of dual numbers $k[\epsilon](\epsilon^2 = 0)$. In short, $(TG)_p = \{ r : k[G] \to k / r$ is k-linear and $r(f.g) = r(f).g(p) + f(p).r(g), f, g \in k[G]\}$. We define $\mathfrak{g}$, the Lie algebra of G, to be the tangent space $(TG)_1$ of G at the identify e. $\mathfrak{g}$ can be canonically given the structure of a 'Lie algebra' (i.e. a bracket operation with some properties). For us, $\mathfrak{g}$ will be regarded as a k-linear space only. Since $k[G]$ is finitely generated, it follows that $\mathfrak{g}$ is finite dimensional. (Let $f_1, \ldots, f_n$ generate $k[G]$ as k-algebra. Define $\phi : \mathfrak{g} \to k^n$ by $\phi(r) = (r(f_1), \ldots, r(f_n))$. Then ϕ is k-linear and injective.)

As an example, consider $G = GL_n$. We have, $k[G] = k[X_{11}, X_{12}, \ldots, X_{nn}]$ - $[\frac{1}{D}]$, D = determinant. For $T \in (TG)_1$, define a matrix $\bar{T} = (\lambda_{ij})$ given by $\lambda_{ij} = T(X_{ij}) \in k$. It is easy to see that this sets up a one-one correspondence between $(TG)_1$ and $M_n(k)$, the algebra of all $n \times n$ matrices. Hence the Lie algebra $\mathfrak{gl}_n$ of GL_n is identified with $M_n(k)$.

Now, the Lie algebra of a principal open subgroup G_1 of G can be identified with that of G itself. The Lie algebra of a closed subgroup is a subspace of the Lie algebra of G. We have a natural action of G on $\mathfrak{g}$, $(\mathrm{Ad}x)(X) = xXx^{-1}$ in any representation, coming from $x(1+\epsilon X)x^{-1} = 1+\epsilon . xXx^{-1}$.

<u>Isogenies.</u> A surjective morphism $f: G \longrightarrow G'$ of connected, reductive algebraic groups G, G' is said to be an <u>isogeny</u> if it has a finite kernel.

It is easy to see that in this case, G is semisimple iff G' is so. Further, we have the following:

(a) Given a semisimple group G', there exists a simply connected (semisimple) algebraic group G and an isogeny $f: G \longrightarrow G'$.

(b) The kernel of any isogeny f is discrete (finite, in fact) and normal. Since G is connected, the kernel of f is central.

(c) f sets up a bijection of conjugacy classes of unipotent elements. (This is because the kernel of f consists of semisimple elements.)

We now prove an important result:

<u>Theorem</u> (Richardson). Let G be a connected algebraic subgroup of GL_n. Let $\mathfrak{g}$ be its Lie algebra ($\mathfrak{g} \hookrightarrow \mathfrak{gl}_n$). Assume that there exists a subspace $\mathfrak{m}$ of $\mathfrak{gl}_n$ such that

$$\left.\begin{array}{ll}(1) & \mathfrak{gl}_n = \mathfrak{g} \oplus \mathfrak{m} \\ (2) & \mathfrak{m} \text{ is stable under } \mathrm{Ad}(G)\end{array}\right\} (*)$$

Then every conjugacy class of GL_n meets G in <u>finitely</u> many conjugacy classes of G.

<u>Proof.</u> Let $G_1 = GL_n$ and $\mathfrak{g}_1 = \mathfrak{gl}_n$. Let C_1 be a conjugacy class of G_1. Consider $C_1 \cap G$. Every conjugacy class of $C_1 \cap G$, being irreducible, is contained in some irreducible component of $C_1 \cap G$ and clearly each component

is a union of classes. Since $C_1 \cap G$ has finitely many components, the theorem is proved if we prove that each component Z of $C_1 \cap G$ is a single class of G. So let Z be such a component. Let C be a class of G such that $C \subseteq Z$. Let $g \in C$. Consider the map $f : G_1 \longrightarrow C_1 g^{-1}$ given by $f(x) = xgx^{-1}g^{-1}$.

Claim: (a) $(df)_1 = 1 - \mathrm{Ad}g$.

(b) $(df)_1$ is surjective.

We have, $(df)_1 : (G_1)_1 \longrightarrow (C_1 g^{-1})_1$. (a) can be verified in a straight forward manner (One may use:

$(1 + \epsilon X).\ g.(1 + \epsilon X)^{-1}.g^{-1} = 1 + \epsilon\,(1 - (\mathrm{Ad}g)(X)))$.

Also, $\dim (df)_1(\mathfrak{g}_1) = \dim \mathfrak{g}_1 - \dim \ker (df)_1$

$= \dim \mathfrak{g}_1 - \dim Z_{\mathfrak{g}_1}(g)$ by (a).

Since $Z_{G_1}(g)$ is an open subset of $Z_{\mathfrak{g}_1}(g)$, consisting of the invertible elements, it has the same dimension.

Hence $\dim (df)_1(\mathfrak{g}_1) = \dim G_1 - \dim Z_{G_1}(g) = \dim C_1 = \dim C_1 g^{-1} = \dim T(C_1 g^{-1})_1$.

From this (b) follows.

$$
\begin{aligned}
\text{Hence}\quad T(Cg^{-1})_1 &\subseteq T(Zg^{-1})_1 \subseteq T(C_1 g^{-1})_1 \cap T(G)_1 \\
&= (1-\mathrm{Ad}g)(\mathfrak{g}_1) \cap \mathfrak{g} \quad \text{(by claim (b) above)} \\
&= (1-\mathrm{Ad}g)(\mathfrak{g}) \quad \text{by } (*) \\
&\subseteq T(Cg^{-1})_1, \quad \text{since } f(G) = Cg^{-1}.
\end{aligned}
$$

It follows that $T(Cg^{-1})_1 = T(Zg^{-1})_1$. Hence $T(C)_g = T(Z)_g$.

Now for a variety V, $\dim (TV)_v \geq \dim V\ \forall v \in V$. The equality holds for almost all $v \in V$ (see the appendix to 2.11). Since C is homogeneous, the equality holds for all elements. Hence $\dim C = \dim T(C)_g\ \forall g \in C$. Also, $\dim T(Z)_g \geq \dim Z$. Thus $\dim C \geq \dim Z$, which gives $\dim C = \dim Z$. Thus C contains a dense open subset of Z. If C' were any other class $\subseteq Z$, then by a similar argument as above, $C' \supseteq$ dense open subset of $Z \Rightarrow C'$ and C intersect, giving $C' = C$. Thus C is the unique class $\subseteq Z$ and hence $C = Z$. This proves the theorem.

Clearly, condition (*) plays an important role in the proof of the above theorem. Hence we try to find out groups for which (*) holds.

Proposition 1. Let G be a group. Then in the following cases, there exists a group G' isogenous to G and a faithful representation $G' \hookrightarrow GL_n$ of it such that (*) holds for $\mathfrak{g}'$, the Lie algebra of G'. In fact, the stronger condition (**) holds: (**) The trace form of $\mathfrak{gl}_n$ is non-degenerate on $\mathfrak{g}'$.

(a) char $k = 0$, G any simple group.

(b) $G = GL_n$.

(c) char $k \neq 2$, G any simple group of type B_n, C_n, D_n.

(d) char $k \neq 2, 3$, G any simple group of type G_2, F_4, E_6, E_7.

(e) char $k \neq 2, 3, 5$, G any simple group of type E_8.

Proof. (a) char $k = 0$. Choose $G' = Ad(G) \subseteq GL(\mathfrak{g})$, where $\mathfrak{g}$ is the Lie algebra of G. Then G' is isogenous to G. (The centre of G is discrete). Also, $\mathfrak{g}'$ is just the Lie algebra ad $\mathfrak{g}$, hence the trace form of $\mathfrak{gl}_n$ on $\mathfrak{g}'$ is just the Killing form of $\mathfrak{g}$ and it is non-degenerate since $\mathfrak{g}$ is simple.

(b) $G = GL_n$; the statement is clear.

(c) Choose the natural representation of G as a classical group (SO_n or Sp_{2n})

Consider SO_n. We claim that its Lie algebra $\mathfrak{g} = \left\{X \in \mathfrak{gl}_n \,\middle|\, X = -X^t\right\}$.
For: O_n is given by functions $\left(\sum_j x_{ij} \cdot x_{kj} - \delta_{ik}\right)_{i,k} \in k[X_{11}, \ldots, X_{nn}]\left[\frac{1}{D}\right]$.
Hence the Lie algebra $\mathfrak{g}$ consists of derivations at I which vanish on these functions.

Hence $T \in \mathfrak{g}$ iff $T(\sum_j x_{ij} . x_{kj} - \delta_{ik}) = 0 \ \forall \ i,k$

$$\text{iff } \sum_j (T(x_{ij}) . \delta_{kj} + \delta_{ij} . T(x_{kj})) = 0 \ \forall \ i,k$$

$$\text{iff } T(x_{ik}) + T(x_{ki}) = 0 \ \forall \ i,k.$$

i.e. The Lie algebra of O_n is the set of all skew symmetric matrices. Since SO_n is the identity component of O_n, its Lie algebra is also the same.

Let $\mathfrak{m} = \{X \in \mathfrak{gl}_n / X = X^t\}$, the spaces of symmetric matrices. Then any $X \in \mathfrak{gl}_n = \frac{(X + X^t)}{2} + \frac{(X - X^t)}{2}$ with $\frac{X + X^t}{2} \in \mathfrak{m}$, $\frac{X - X^t}{2} \in \mathfrak{g}$.

Further, for $A \in \mathfrak{g}$, $B \in \mathfrak{m}$, $tr(AB) = tr((AB)^t) = tr(-BA) = -tr(AB)$. This shows that $tr(AB) = 0$. Hence $\mathfrak{gl}_n = \mathfrak{g} \oplus \mathfrak{m}$. Since the trace form is non-degenerate on $\mathfrak{gl}_n$ and $\mathfrak{g}, \mathfrak{m}$ are orthogonal, (**) follows.

Consider $Sp_{2n} = \{A \in SL_{2n} / M(A^t)^{-1} M^{-1} = A\}$, $M = \begin{bmatrix} 0 & & & & & +1 \\ & & & & +1 & \\ & & & +1 & & \\ & & -1 & & & \\ & -1 & & & & \\ -1 & & & & & 0 \end{bmatrix}_{2n}$

We claim that its Lie algebra $\mathfrak{g}$ is given by: $\mathfrak{g} = \{X \in \mathfrak{gl}_{2n} / XM + MX^t = 0\}$.

(This claim is easily proved in the same way as in the case of SO_n.) We note that $X \in \mathfrak{g}$ iff XM is symmetric. Let $\mathfrak{m} = \{X \in \mathfrak{gl}_{2n} / XM \text{ is skew-symmetric}\}$.

It immediately follows that $\mathfrak{gl}_{2n} = \mathfrak{g} + \mathfrak{m}$. Again, for $A \in \mathfrak{g}, B \in \mathfrak{m}$, $A = MA^tM$; $B = -MB^tM$.

Also, $tr(AB) = tr(MA^t.M. (-MB^tM)) = tr(M.A^tB^tM)$ as $M^2 = -I$

$$= -tr(A^tB^t) \text{ as can easily be verified}$$

$$= -tr(BA)^t = -tr(BA) = -tr(AB).$$

Hence $tr(AB) = 0$ since char $k \neq 2$. Hence $\mathfrak{gl}_{2n} = \mathfrak{g} \oplus \mathfrak{m}$ and the result follows.

For (d) and (e), we again choose the adjoint representation. We observe that the Lie algebra of a simple group possesses a special basis, called Chevalley basis'. We calculate the discriminant of the Killing form with respect to this. This is a number (integer) which is divisible only by 2, 3 for the groups in (d) and by 2, 3, 5 for the group in (e). Hence (d) and (e) hold if we put the suitable restrictions on char k.

Definition. Given a root system (of a reductive group G), a prime p is said to be 'good' with respect to it if p satisfies:

(1) Root system simple, and of type:

A_n : p arbitrary

B_n, C_n, D_n : $p \neq 2$

G_2, F_4, E_6, E_7 : $p \neq 2, 3$,

E_8 : $p \neq 2, 3, 5$.

(2) Root system is not simple. Let $R = R_1 \cup \ldots \cup R_k$ be the simple components; then p is good with respect to each of R_i's (as defined in (1)).

Remark. The property 'p good' is inherited by integrally closed subsystems.

Theorem 2. If G is reductive and char G (= char k) is good (with respect to the root system of G), then the number of unipotent conjugacy classes is finite.

Proof. Let $G = GL_n$ (or SL_n). Every unipotent element is of the form: $g.A.g^{-1}$, where A is of the type: $\begin{bmatrix} A_1 & 0 & \ldots & 0 \\ 0 & A_2 & \ldots & 0 \\ \cdot & & \ldots & \\ \cdot & & \ldots & A_k \end{bmatrix}$ where each A_i is of the form : $\begin{bmatrix} 1 & 1 & 0 & 0 \ldots \\ 0 & 1 & 1 & 0 \ldots \\ \cdot & \cdot & \cdot & \ldots \\ 0 & \cdot & \cdot & \ldots 1 \end{bmatrix}$ $1 \leq i \leq k$. (The Jordan normal form). Let $r_1, \ldots, r_k$

be the 'block - sizes'. Then $r_1, \ldots, r_k$ completely determine the conjugacy class (of unipotent elements) to which A belongs. In other words, the number of distinct conjugacy classes = the number of collections of integers $(r_1, \ldots, r_n)$ such that $r_i \geqslant 0$ and $\sum r_i = n$ (= p(n), the number of partitions of n into n non-negative integers). Hence the theorem is true for GL_n or SL_n (irrespective of char G).

Let G be any arbitrary semisimple group. Then there exists an isogeny $f : G' \longrightarrow G$, where G' is simply connected. Now, the number of unipotent conjugacy classes of G = that of G'. Hence we may prove the theorem for G'. In other words, we may assume G to be simply connected. G, being semisimple and simply connected, is a finite direct product of simple groups G_i. Hence the theorem need be proved only for simple groups G_i, which by the above may be assumed to be different from A_r. Since char G is good, it follows that char G_i is also good. (The root system of G_i is an integrally closed subsystem of that of G). Now the proof of the proposition 1 shows that the condition (*) is satisfied for G_i. (In fact, (**) is satisfied, by taking a suitable isogenous group). Hence by Richardson's theorem, any class of GL_{n_i}, ($\rho_i : G_i \longrightarrow GL_{n_i}$ is a faithful representation) meets G_i in finitely many classes. Since GL_{n_i} itself has finitely many unipotent conjugacy classes, it follows that G also has this property.

This proves the theorem.

Remark. It is not known whether the hypothesis on char G is necessary or not.

Corollary 1. In a reductive group G, with char G good, the number of conjugacy classes of centralizers of elements of G is finite.

Proof. Let T be a maximal torus of G. Then the number of centralizers, in G, of elements of T is finite (by corollary to proposition of 2.7). Since any semisimple element is conjugate to an element in T, it follows that the number of conjugacy classes of centralizers of semisimple elements is finite. Let $x \in G$, $x = s.u$ be the Jordan decomposition. Then $Z_G(x) = Z_G(s) \cap Z_G(u)$. (Up to conjugacy, there are finitely many possibilities for $Z_G(s)$.) Again, $u \in Z_G(s)^o$ (by corollary 4 to theorem 1 of 2.13). $Z_G(s)^o$ is reductive (by proposition 4 of 3.5) and char $(Z_G(s)^o)$ can be seen to be good. Hence up to conjugacy, u has finitely many possibilities in $Z_G(s)^o$. Hence $Z_G(x)$ itself has finitely many possibilities in G. This proves the corollary.

Remark. By using a similar method, one can prove the following: Let G be a connected, reductive group with char $G = 0$ or sufficiently large. Let G act on an affine variety V. Then the number of conjugacy classes of Levi components of G_v $(v \in V)$ is finite. (If we write $G_v = M.U$, semidirect, with U the unipotent radical and M reductive, then M is called a Levi component.)

Theorem 3. Let G be a reductive group with char G good (or with the property: G has finitely many unipotent conjugacy classes). Let V be the set of all unipotent elements in G. Then,

(a) V is a closed, irreducible subvariety of G and it has codimension r in G (r = rank of G).

(b) V contains a unique class of regular elements. (Thus, in particular, regular unipotent elements exist in case char G is good). This class is open dense in V and its complement has codimension $\geqslant 2$ in V.

<u>Proof.</u> Take a faithful representation in GL_n of G. Now the unipotent matrices in GL_n form a closed set (A is unipotent iff $(A-I)^n = 0$, a polynomial condition). It follows that V is also closed in G. (This, of course, is true if G is any group). Fix a Borel subgroup B in G.

Define $S \subseteq G/B \times G$ by : $S = \{(gB, x)/g^{-1}xg \in U\}$, $U = B_u$. It is clear that S is well defined and closed. Also, S is the image of $G \times U$ under the morphism $\phi : G \times U \longrightarrow G/B \times G$ given by : $\phi(g, x) = (gB, gxg^{-1})$. Hence S is irreducible. Consider the projection p_1 of $G/B \times G$ onto the first factor. Clearly $p_1(S) = G/B$. Also, the fibres of p_1 are conjugates of U, hence are of the same dimension.

Hence $\dim S = \dim S = \dim G/B + \dim U$

$= \dim G - \dim B + \dim U$

$= \dim G - r$ (since $\dim B - \dim U = r$).

Consider the projection p_2 onto the second factor. Clearly $p_2(S) = V$. Now we show that some fibre of p_2 is finite. This proves that $\dim S = \dim p_2(S) = \dim V$. Hence V has codimension r in G. Choose $x = \prod_{\alpha > 0} x_\alpha(c_\alpha) \in U$ such that $c_\alpha \neq 0$ for all simple roots. We show : $g^{-1}xg \in U \Rightarrow g \in B$. This will clearly prove that the fibre of p_2 over x is finite, in fact consists of only one element viz. (B, x).

As seen in proposition 4 of 3.5, $g^{-1} = u.n_w.b$, $u \in U_w$. One can assume $b = 1$. Now we have : $g^{-1}\ xg \in U$ i.e. $un_w.x.n_w^{-1}u^{-1} \in U$ i.e. $n_w x n_w^{-1} \in U$. Now $n_w.x.n_w^{-1} \in \prod_{\alpha > 0} X_{w(\alpha)}$ also. Hence $w(\alpha) > 0$ whenever $c_\alpha \neq 0$. In particular, $w(\alpha) > 0$ <u>for all simple roots</u> α. This clearly means $w = \mathrm{Id}$. (w takes the fundamental chamber into itself). Hence $n_w \in T$, so that $g \in B$.

This proves (a). (The argument above proves, incidentally, that $gBg^{-1} = B \Rightarrow g \in B$ (for reductive groups)).

(b) Since G has only finitely many unipotent conjugacy classes, V is a <u>finite</u>

union of conjugacy classes. Hence dimension of V is equal to the dimension of at least one of the classes. Since dim V = dim G - r, it follows that some class in V also has dimension equal to dim G - r. Hence this class C_o is regular. Also, its closure is the whole of V. Since any class is open in its closure, it follows that C_o is open (dense) in V. Now any other class in V is of strictly lower dimension (corollary 4 to proposition 1 of 1.13) and hence cannot be regular. This proves (b). The statement about codim is proved later.

Remarks. (1) The conclusions (a) and (b) hold in arbitrary characteristics and we shall use them. (All we require is the existance of regular unipotent elements). (2) The set of irregular unipotent elements is closed in G (and has codimension ≥ 2).

3.7 Regular elements (continued).

We now characterize the regular unipotent elements.

Theorem 1. Let G be a reductive group, T a maximal torus and B = T. U, a Borel subgroup containing T. Let x be a unipotent element in G. Then the following statements are equivalent.

(a) x is regular.

(b) x belongs to a unique Borel subgroup.

(c) x belongs to finitely many Borel subgroups.

(d) If $x \in U$, $x = \prod_{\alpha > 0} x_\alpha(c_\alpha)$, then $c_\alpha \neq 0$ for every simple root α.

Proof. Let $G = T_o \cdot S$, where T_o is the radical and $S = [G,G]$, a semisimple group. T_o is a torus and it is central. T_o is contained in every Borel subgroup.

If $y = t_0.y'$ with $t_0 \in T_0$, $y' \in S$, then y is regular in G iff y' is regular in S. It is now clear that in proving the equivalence of the above statements, one may assume G itself to be semisimple.

(b) $\Longrightarrow$ (c) is obvious.

(c) $\Longrightarrow$ (d). Let $x = \prod_{\alpha > 0} x_\alpha(c_\alpha)$. If possible, let $c_{\alpha_0} = 0$ for some simple root α_0. Now $\{\alpha > 0, \alpha \neq \alpha_0\} \cup \{-\alpha_0\}$ is again a set of positive roots (with respect to some ordering, since w_{α_0} permutes positive roots $\neq \alpha_0$). Hence $X_{-\alpha_0}$ normalizes $U_{\alpha_o} = \prod_{\substack{\alpha > 0 \\ \alpha \neq \alpha_o}} X_\alpha$. It follows that $x \in {}^yB(y \in U_{-\alpha_0})$, which are infinitely many in number. (For $d \neq 0$, $x_{-\alpha_0}(d) \notin B$.) This contradicts our assumption in (c) and hence proves (d).

(d) $\Longrightarrow$ (a). Clearly, the elements satisfying the condition (d) are dense in V. By theorem 3, regular unipotent elements are also dense in V. Hence $\exists\, x_0 \in U$ such that x_0 is regular and satisfies (d). Our claim is that x and x_0 are conjugate (in B) and this proves the implication.

For the proof we develope here some machinery which will be useful in later discussions also.

(1) <u>Commutation Formulae:</u>

For positive roots α, β, $(x_\alpha(t), x_\beta(u)) = \prod_{r > 0} x_r(p_r(t,u))$, where $p_r(t,u)$ is a polynomial in t,u. This polynomial is identically zero if $r \neq i\alpha + j\beta$ (i,j integers $\geqslant 1$) and $p_{i\alpha + j\beta}(t,u) = c(i,j)\, t^i.u^j$, $c(i,j) \in k$.

This can be easily proved in the same manner as the proof of proposition 2 of 3.3.

(2) For a simple root α_i, let $U_i = \prod_{\substack{\alpha > 0 \\ \alpha \neq \alpha_i}} X_\alpha$. Let $U' = \bigcap_{1 \leq i \leq r} U_i$. Then the

above formulae show : $U' \supseteq [U,U]$, hence U' is normal in U and U/U' is abelian. Also, we note that codim U'(in U) = r.

Coming back to the claim, let $x_0 = \prod_{\alpha>0} x_\alpha(c_\alpha)$ and $x = \prod_{\alpha>0} x_\alpha(d_\alpha)$ with $c_\alpha \neq 0$, $d_\alpha \neq 0$ for every simple root α. By proposition 7 of 2.6, choose $t \in T$ such that $\alpha(t).d_\alpha = c_\alpha$ for <u>all simple</u> roots α.

Hence $x' = txt^{-1} = \prod_{\alpha>0} x_\alpha(\alpha(t).d_\alpha)$.

Now it is enough to show that x' and x_0 are conjugate. Clearly $x'x_0^{-1} \in U'$.

Since $C_U(x_0).x_0^{-1} \subseteq U'$, $\dim C_U(x_0).\ x_0^{-1} \leq \dim U' = \dim U - r$. Also, $\dim Z_U(x_0) \leq \dim Z_G(x_0) = r$, since x_0 is regular.

Thus $\dim C_U(x_0) \geq \dim U - r$. Thus $\dim C_U(x_0)\ .\ x_0^{-1} = \dim U'$. Also, $C_U(x_0)$ is closed, being a class in <u>unipotent</u> group (by corollary to proposition of 2.5). Since U' is irreducible and $C_U(x_0).x_0^{-1}$ is closed and has the same dimension as U', it follows that $C_U(x_0)\ .x_0^{-1} = U'$.

We have: $x'x_0^{-1} \in U'$. Hence $\exists\, u \in U$ such that $ux_0u^{-1}x_0^{-1} = x'x_0^{-1}$ or $x' = ux_0u^{-1}$. This proves the claim.

(a) $\Rightarrow$ (b). Let x be regular. Choose a $x_0 \in U$ such that x_0 satisfies the condition (d). Hence x_0 is regular (as proved earlier). However, by theorem 3 of 3.6, x and x_0 are conjugate. Also, the proof of this theorem 3 shows that x_0 is contained in a unique Borel subgroup. It follows that x is also contained in a unique Borel subgroup.

This proves the theorem completely.

<u>Corollary</u>. If x is regular, $x \in U$, then $Z_G(x)^o \subseteq U$ and hence is unipotent. (For: In the proof of (d) $\Rightarrow$ (a), the inequalities all become equalities and hence $\dim Z_G(x_0)^o = \dim Z_U(x_0)^o$.)

We again consider U_i as defined earlier. Let $P_i = (T.\langle X_{\alpha_i}, X_{-\alpha_i}\rangle).U_i$, where $(\langle X_{\alpha_i}, X_{-\alpha_i}\rangle)$ is the group generated by X_{α_i} and $X_{-\alpha_i}$. Both T and $\langle X_{\alpha_i}, X_{-\alpha_i}\rangle$ normalize U_i (because of the commutation formulae). Here, P_i is a rank 1 - parabolic subgroup, U_i is its unipotent radical and $T.\langle X_{\alpha_i}, X_{-\alpha_i}\rangle$ is a Levi component of P_i.

Now, $\dim T.\langle X_{\alpha_i}, X_{-\alpha_i}\rangle = r+2$. (The overlap of T and $\langle X_{\alpha_i}, X_{-\alpha_i}\rangle$ has dimension 1). Hence $\dim P_i - \dim U_i = r + 2$.

Now, $P_i \supseteq B$ and hence G/P_i is complete. Consider the set $S_i \subseteq G/P_i \times G$, given by : $S_i = \left\{(gP_i, x) / g^{-1}xg \in U_i\right\}$ (which is well defined). Then S_i is closed and irreducible (being image of the morphism $G \times U_i \xrightarrow{\phi} G/P_i \times G$, $\phi(g,x) = (gP_i, gxg^{-1})$). By projecting onto the first factor,

$\dim S_i = \dim G/P_i + \dim U_i$ (by an argument similar to one in theorem 3) = $\dim G - (r+2)$.

By taking projection onto the second factor, $p_2(S_i) = V_i = \bigcup_{g \in G} gU_ig^{-1}$ and $\dim V_i \leq \dim S_i = \dim G - (r+2)$.

Now, by the proposition above, an unipotent element x is irregular iff $x \in V_i$ for some i. Hence, from above,

dim (V - regular uni. elements) = $\sup_i \dim V_i \leq \dim G - (r+2)$.

This proves the unproved part of statement (b) of theorem 3.

Incidently, each V_i defined above is closed, since G/P_i is complete. This shows once again that the set of irregular unipotent elements is closed.

This argument can be carried out in case of arbitrary irregular elements of G. We start with a lemma:

Lemma 1. For each α_i, let $T_i = \ker \alpha_i$. Let U_i, P_i be as before, $B_i = T_i.U_i$ (B_i has codimension 2 in B). Then every irregular element of G is contained in a conjugate of some B_i. Converse is also true.

For the proof, see I.H.E.S. # 25 (§ 5).

From this, we get: The irregular elements of G form a closed subset of G, each of whose components has codimension 3. In particular, regular elements form a dense open subset of G. The proof of this statement is like that just given, but with B_i in place of U_i. We have $\operatorname{codim}_{P_i} B_i = 3$.

Lemma 2. Among the irregular elements, the semisimple ones are dense.

Proof. For each i, set $J_i = \ker \alpha_i - \bigcup_{\alpha > 0, \alpha \neq \alpha_i} \ker \alpha$. Consider $J_i . U_i$, which is open in B_i. Our claim is that $J_i . U_i$ consists of semisimple elements. So let $x = t.u \in J_i . U_i$, $t \in J_i, u \in U_i$. Now by conjugating by a suitable element b_0 of B_i, we have: $x' = b_0 x b_0^{-1} = t.u'$ with $u' \in U_i$ and $t.u' = u'.t$. It is now easy to see that $u' = 1$ so that x' is semisimple. Hence x is also semisimple. Since $\bigcup J_i U_i$ is dense in $\bigcup B_i$, $\bigcup_i \bigcup_g g.J_i.U_i.g^{-1}$ is dense in $\bigcup_{g,i} gB_i g^{-1} = Ir$, the set of irregular elements. This proves the lemma.

We now characterize the regular elements of G in another way.

Theorem 2. Let x be any element of G (G reductive). Then x is regular iff the number of Borel subgroups containing x is finite. (This number is 1 if x is unipotent and equal to $|W|$ if x semisimple).

For the proof, we require a lemma.

Lemma 3. Let $t \in G$ be semisimple. Set $G_0 = Z_G(t)^0$. Then each Borel subgroup of G, containing t, contains a unique Borel subgroup of G_0.

Conversely, each Borel subgroup of G_0 contains t and is contained in finitely many Borel subgroups of G.

Proof. Let B be a Borel subgroup of G, containing t. Then from the proof of proposition 4 of 3.5, we see that $B \cap G_0$ (the group generated by T and all X'_αs such that $\alpha > 0, \alpha(t) = 1$) is a Borel subgroup of G_0, clearly the unique one contained in B. Conversely, let B_0 be a Borel subgroup of G_0. It contains t since t is in the centre of G_0. Let T_0 be a maximal torus of G_0, contained in B_0. It is also maximal in G since rank G = rank G_0. As such it is contained in exactly $|W|$ Borel subgroups of G. At most $|W|$ Borel subgroups of G can thus contain B_0. This proves the lemma.

Coming back to the proof of the theorem, let $x = t.u$ be the Jordan decomposition of x. It follows that for a Borel subgroup B of G, $x \in B$ iff both t and u are in B. Also, by proposition 5 of 3.5, x is regular in G iff u is regular in $Z_G(t)^0 = G_0$ say. The proof is now immediate from the above lemma and the theorem.

Remark. If W_0 denotes the Weyl group of G_0, then a closer analysis shows that the finite number in the lemma is $|W| / |W_0|$ (and accordingly in the theorem with $t = x_s$).

Theorem 3. If G is a reductive group, then the map $x \rightsquigarrow x_s$ yields a bijection of the regular classes onto the semisimple classes.

Proof. (1) Surjectivity. Let s be a semisimple element. Then by proposition 4 of 3.5, $Z_G(s)^0$ is reductive. Hence by theorem 3 above, $\exists$ $u \in Z_G(s)^0$ which is regular, unipotent. Now $x = s.u$ is the Jordan decomposition to x and by proposition 5 of 3.5, x is regular since u is so in $Z_G(s)^o$. Also, $x_s = s$. This proves the surjectivity.

(2) Injectivity. Let x, y be regular elements such that x_s and y_s are conjugate. By conjugating x by a suitable element (in fact by g, where $gx_sg^{-1} = y_s$), one may assume, without loss of generality, that $x_s = y_s = s$. Consider $Z_G(s)^o$, which is reductive. Since x_u, y_u are regular, unipotent elements in $Z_G(s)^o$, it follows that x_u, y_u are conjugate in $Z_G(s)^o$ (Theorem 3). So let $g.x_ug^{-1} = y_u$, $g \in Z_G(s)^o$.

Now $gxg^{-1} = g.x_s.x_ug^{-1} = g.sx_ug^{-1} = s.gx_ug^{-1} = s.y_u = y$.

This proves the injectivity.

3.8. Regular elements in simply connected, semisimple groups.

In this section, G is assumed to be a simply connected semisimple algebraic group.

Let T be a fixed maximal torus and R be the root system relative to it. We recall that the fundamental weights $\{\lambda_i\}_{1 \leq i \leq n}$ in this case, are in fact characters (by definition of simply connectedness). We denote the corresponding characters on G by X_i ($X_i = X_{\lambda_i}$). Consider the map $p : G \rightarrow \mathbb{A}^r$ given by $p(g) = (X_1(g), \ldots, X_r(g))$. Then we have:

Theorem 1. Let F be any fibre of p.

(a) F is a closed, irreducible subvariety (of G) which has codimension r in G.

(b) F is a union of conjugacy classes, finite in number in case of good characteristics.

(c) F contains a unique class of regular elements. This class is open and dense in F and its complement has codimension ≥ 2 (in F).

(d) F contains a unique class of semisimple elements. This class is closed, has the minimal dimension among the classes in F, is in the closure of any of the classes in F, and is characterized by any of these properties.

Proof. Since there always exists $t \in T$ such that $X_i(t) = c_i \ \forall i$, for a pregiven r-tuple $(c_1, \ldots, c_r)$, it follows that F is always non-empty. Clearly F is closed and is a union of conjugacy classes (since X_i's are class functions). Also, F contains a semisimple class S. Because of corollary 3 to theorem 2 of 3.4, this class is unique. By proposition 4 of 3.6, pick the regular class C which corresponds to S. Since $X_i(x) = X_i(x_s)$, it follows that $C \subseteq F$ as well and is unique.

We claim that C is dense in F. So let $y \in F$ be arbitrary and H, open in F, contain y. Let H', open in G, be such that $H' \cap F = H$. Consider $Z_G(y_s)^o$. Now, $y_u \in y_s^{-1} \cdot H' \cap V$, where V is the set of all unipotent elements of the reductive group $Z_G(y_s)^o$. Now by theorem 3 of 3.6, the class of all regular unipotent elements in $Z_G(y_s)^o$ is dense in V. Hence $\exists$ u, regular unipotent $\in Z_G(y_s)^o$, which is also in $y_s^{-1} H' \cap V$. Hence $x = y_s . u \in H'$ and $x_u = u$. Since $X_i(x) = X_i(y_s) \ \forall$ i, it follows that $x \in F$. Since $x_u = u$ is regular in $Z_G(y_s)^o$, so is $x \in G$. Hence $x \in C$ and also in $H' \cap F = H$. This proves the claim.

Now C is irreducible (being image of G under the conjugation map $g \rightsquigarrow g x_0 g^{-1}$, $x_0 \in C$ fixed). Hence F itself is irreducible. Now F is union of C and other classes of strictly lower dimension. Hence $\dim F = \dim C$. C is open in F as well, since $\overline{C} = F$. But then C is regular and has codimension $= r$. Hence F also has codimension r. The fact that complement of C

has codim ≥ 2, is derived easily from a similar statement in theorem 3 of 3.6. This proves (a), (b) and (c).

(d) For any class $S_1 \subseteq F$, we prove the equivalence of the following statements:

(i) S_1 is the (unique) semisimple class S.

(ii) S_1 is closed.

(iii) S_1 has minimal dimension among classes in F.

(iv) S_1 belongs to the closure of any of the classes of F.

(iv) $\Rightarrow$ (iii) $\Rightarrow$ (ii). These follow immediately from corollary 4 to proposition 1 of 1.13. (Note that F is closed.)

(ii) $\Rightarrow$ (i) is already proved in corollary 5 to theorem 2 of 3.4.

(i) $\Rightarrow$ (iv) follows from Lemma in 3.4 and uniqueness of S. (Take any class K in F, take $x \in K$, then $x_s \in S$ also, $x_s \in \mathrm{cl}\, K \Rightarrow S \subseteq \mathrm{cl}\, K$).

This proves the theorem completely.

Theorem 2. The regular classes have a natural structure of a variety, isomorphic to $\mathbb{A}^r$, under the map $p : G^{reg.} \longrightarrow \mathbb{A}^r$, where G^{reg} is the open variety of G of regular elements.

The points to be proved are:

(1) p is a morphism and its fibres are just the (regular) classes.

(2) $p^*(k[\mathbb{A}^r]) = k[G^{reg}]^{\mathrm{Int}\, G}$.

(3) If $f \in k[\mathbb{A}^r]$, $x \in G^{reg}$, then f is <u>defined</u> at $p(x)$ iff $p^*(f)$ is defined at x.

The proof of these points is straight forward and is ommitted from here. (e.g. x, y regular and $p(x) = p(y) \Leftrightarrow p(x_s) = p(y_s) \Leftrightarrow x_s$ conjugate to $y_s \Leftrightarrow x$ conjugate to y).

We now give a final important characterization of regular elements (in case of simply connected group, of course).

Theorem 3. Let $x \in G$, $p : G \to A^r$ as before. Then x is regular iff $(dp)_x$ is surjective. i.e. iff $(dX_i)_x$ $(1 \leq i \leq r)$ are linearly independent.

We postpone the proof of this theorem for a while and give a development which will eventually prove it and at the same time will produce a cross-section to the collection of regular classes.

Cross-Sections. Let $G, T, \alpha_1, \ldots, \alpha_r, X_1, \ldots, X_r$ be as before. Pick $n_i \in N(T)$, a representative for w_i = reflection relating to α_i.

Consider $X_{\alpha_1} \cdot n_1 X_{\alpha_2} \cdot n_2 \ldots X_{\alpha_r} \cdot n_r = C$. We shall show that C is a cross-section of the regular classes.

We have, $n_1 X_{\alpha_2} = X_{w_1(\alpha_2)} \cdot n_1$; $n_1 n_2 X_{\alpha_3} = X_{w_1 w_2(\alpha_3)} \cdot n_1 n_2$ etc.

Proceeding in this way, we get, $C = X_{\beta_1} \cdot X_{\beta_2} \ldots X_{\beta_r} \cdot n_1 \ldots n_r$, where $\beta_i = (w_1 \ldots w_{i-1}) \cdot (\alpha_i)$, $1 \leq i \leq r$.

Now the following facts about these β_i's can be verified easily:

(1) $\beta_1, \ldots, \beta_r$ are just the positive roots which are made negative by $w^{-1} = w_r^{-1} \ldots w_1^{-1}$. (This follows easily from the fact: w_i permutes all the positive roots other than α_i.)

(2) $\beta_1, \ldots, \beta_r$ are linearly independent (for $\beta_i = \alpha_i +$ earlier α_j's). Hence, as a consequence of (1), $s\beta_i + t\beta_j$ is not a root for $i \neq j$, s and t integers ≥ 1.

(3) As a consequence of the commutation formulae in proposition 2 of 3.6, it follows that X_{β_i} and X_{β_j} commute elementwise.
Thus C may be written $U_w . n_w$ with $w = w_1 \dots w_r$, the translation by n_w of an affine r-dimensional space.

We now aim to prove:

Theorem 4. Let $C = \prod_{i=1}^{r} X_{\alpha_i} n_i$ be as above.
(a) C is a closed subset of G and isomorphic, as a variety, to $\mathbb{A}^r$, the co-ordinates coming from those of X'_{α_i} s.
(b) C is a cross-section of the collection of regular classes.

Here (a) follows from the above discussion.
To prove (b) we require:

Theorem 5. $p : C \longrightarrow \mathbb{A}^r$ is an isomorphism of varieties.

Before proving this theorem, we consider the example $\underline{G = SL_n}$. The simple roots are $\{\alpha(i,i+1)\}_{1 \leq i \leq n-1}$ and the corresponding unipotent groups are $X_i = \{I + t_i . E_{i,i+1} \mid t_i \in k\}$. For n_i, we choose a monomial matrix as below.
If $n = 2$, then an element g of C is of the form: $\begin{bmatrix} 1 & t \\ 0 & 1 \end{bmatrix} . \begin{bmatrix} 0 & -1 \\ 1 & 0 \end{bmatrix} = \begin{bmatrix} t & -1 \\ 1 & 0 \end{bmatrix}$.
It is now clear that for arbitrary n, an element g of C is of the form:

$$\left[\begin{array}{cc|c} t_1 & -1 & 0 \\ 1 & 0 & \\ \hline & 0 & I \end{array}\right] . \left[\begin{array}{c|cc|c} 1 & 0 & & \\ \hline 0 & t_2 & -1 & 0 \\ & 1 & 0 & \\ \hline & 0 & & I \end{array}\right] \dots\dots$$

$$= \left[\begin{array}{c|c} t_1 \ -t_2 \dots & (-1)^{n-1} \\ \hline I_{n-1} & 0 \end{array}\right] = \underbrace{\left[\begin{array}{c|c} 1 & t_1 \ -t_2 \dots \\ \hline 0 & I_{n-1} \end{array}\right]}_{u_w} . \underbrace{\left[\begin{array}{c|c} 0 \dots 0 & (-1)^{n-1} \\ \hline I_{n-1} & 0 \end{array}\right]}_{n_w}, \text{ i.e.}$$

it is in Jordan normal form (one block - see proposition 2(c) of 3.5)). Also, for any $k (1 \leq k \leq n-1)$, $SL_n \to \mathrm{End}(\Lambda^k(k^n))$ is the irreducible representation for the fundamental weight λ_k. It is easy to see that $X_k(g) = \sum$ principal k-minors. For $g \in C$, the only nonzero principal k-minor that appears is that in the upper left corner and in this case the contribution is t_k. Hence $X_k = t_k$, and Theorem 5 is verified in this case. The proof in the general case is similar, as we now show.

To prove theorem 5, we require the following development:

Definition. Let $\lambda_1, \ldots, \lambda_r$ be the fundamental weights. Define $\lambda_i \dot{>} \lambda_j$ if there exists a dominant weight λ such that $\lambda_i \geqslant \lambda$ (i.e. $\lambda_i - \lambda$ is a sum of positive roots) and λ_j is in the support of λ. (Supp $\lambda = \{\lambda_k$ such that $\lambda = \sum_i n_i \lambda_i$ with $n_k \neq 0\}$).

It can easily be checked that this is indeed a well-defined (strict) partial order. Sometimes, e.g. for type A_r, it is vacuous.

Main lemma. Let $y \in C$, $y = y_1 \ldots y_r$, $y_i = x_{\alpha_i}(t_i).n_i$. Then for each i, $X_i(y) = c_i . t_i$ + polynomial in earlier (with respect to $\dot{>}$) t_j's, where c_i is a non-zero constant, independent of t_i's.

Proof. Let $V = V_i^{\rho_i}$ be the irreducible representation space of G with λ_i as the highest weight. Then $X_i(g) = Tr_V(g)$, $g \in G$.
Now, by the fundamental theorem of representation theory, $V = \bigoplus V_\lambda$, where each λ is of the form $\lambda_i - \sum_k n_k \alpha_k$ with $n_k \geqslant 0$. Also, V_{λ_i} is of dimension 1. Let $0 \neq v \in V_\lambda$, where $\lambda = \sum_j m_j \lambda_j$, $m_j \in \mathbb{Z}$. By the lemma of 3.3,

$n_j.v \in V_{w_j.\lambda} = V_{\lambda - m_j\alpha_j}$. Hence by proposition 2 of 3.3, $x_{\alpha_j}(c).n_j.v = \sum_{m \geq 0} c^m.v_m$ (finite sum) with $v_m \in V_{\lambda - m_j\alpha_j + m\alpha_j}$, independent of c and $v_0 = n_j.v$.

It follows that if $y = y_1 \ldots y_r, y_j = x_{\alpha_j}(t_j).n_j$, then

$$y.v = \sum t_1^{k_1} \ldots t_r^{k_r}.v_{(k_1,\ldots,k_r)} \qquad (**)$$

where $k_i \geq 0 \ \forall i$, $v_{(k_1,\ldots,k_r)} \in V_\mu$ with $\mu = \lambda + \sum_j (k_j - m_j)\alpha_j$.

Consider $Tr_V(y) = \sum_\lambda Tr_{V_\lambda} \pi_\lambda.y.n_\lambda$, where $\pi_\lambda : V \longrightarrow V_\lambda$ is the projection and $n_\lambda : V_\lambda \hookrightarrow V$ is the injection. From (**) above, $y.v$ contributes to the trace only if $k_j = m_j \ \forall j$.

Hence:

(1) The contribution is zero if $m_j < 0$ for some j i.e. if λ is not a dominant weight: for the right side of (**) has no term back in V_λ.

(2) If λ is dominant $\neq \lambda_i$ and $m_j = 0$ for some j, then t_j does not occur in the contribution. In other words, only those t_j's may occur where $m_j \neq 0$. But then in that case $\lambda_i \gneq \lambda_j$. Hence the contribution is a polynomial in earlier t_j's.

(3) If $\lambda = \lambda_i$, then $m_j = 0 \ \forall j \neq i$, $m_i = 1$. Hence the contribution is $c_i t_i$ with c_i independent of the c_j's, by (**). We claim that $c_i \neq 0$. For this we observe that each y_j, $j \neq i$, acts as a nonzero scalar on V_{λ_i} (scalar by (**), nonzero since also $w_j\lambda_i = \lambda_i$ and $\dim V_{\lambda_i} = 1$). Hence $c_i = 0$ gives $X_{\alpha_i} n_i V_{\lambda_i} = V_{\lambda_i - \alpha_i}$, thus $X_{\alpha_i} V_{\lambda_i - \alpha_i} = V_{\lambda_i - \alpha_i}$. But also $X_{-\alpha_i} V_{\lambda_i - \alpha_i} = X_{\lambda_i - \alpha_i}$ (since $\lambda_i + \alpha_i$ is not a weight on V, neither is $w_i(\lambda_i + \alpha_i) = \lambda_i - 2\alpha_i$). Hence $n_i V_{\lambda_i - \alpha_i} = V_{\lambda_i - \alpha_i}$ since $n_i \in \langle X_{\alpha_i}, X_{-\alpha_i} \rangle$. This gives

a contradiction since $n_i(V_{\lambda_i}) = V_{\lambda_i - \alpha_i}$. This clearly proves the result.

Remark. By choosing n_j's properly, it is possible to prove:
$X_i(y) = t_i +$ polynomial$/\mathbb{Z}$ in earlier t_j's.

Proof of theorem 5. From the above lemma, it follows that each t_i can written in the form: $X_i(y)$ + polynomial in earlier $X_j(y)'$s. In view of the part (a) of of theorem 4, it follows immediately that $p: C \longrightarrow \mathbb{A}^r$ is an isomorphism of varieties.

We now prove theorem 3 and part (b) of theorem 4 simultaneously by proving:

Theorem 6. Let $x \in G$, $C \subseteq G$ as above. Then the following conditions are equivalent:

(1) x is regular

(2) $(dp)_x$ is surjective.

(3) x is conjugate to some element in C.

(1) $\Longleftrightarrow$ (2) is the theorem 3 and by (3), C consists of regular elements, one from each class (Theorem 1 and 5).

Proof. (3) $\Longrightarrow$ (2). Clearly, one may assume $x \in C$. Now by theorem 5, $(dp)_x : (TC)_x \longrightarrow (T\mathbb{A}^r)_{p(x)}$ is surjective. Also, $(TC)_x$ is a subspace of $(TG)_x$. Hence $(dp)_x$ is surjective.

(2) $\Longrightarrow$ (1). We prove: x is irregular $\Longrightarrow dX_1, \ldots, dX_r$ are linearly independent at x (i.e. equivalently $(dp)_x$ is not surjective).

Step I. We need prove the above statement for semisimple elements only.

For: By lemma 2 of 3.7, the set of semisimple, irregular elements is dense in the set of all irregular elements. Hence our claim is clear if the set of irregular elements at which $dX_1, \ldots, dX_r$ are linearly dependent is <u>closed</u>. Now choose a basis to the tangent space at e. Then the left translation by y gives a basis to the tangent space at y. Consider the dual base. dX_i is a linear combination of the basis elements. Now from the choice of the basis, it is clear that dX_i is a vector field on G (i.e. the coefficients polynomial functions). Consider the matrix formed by these coefficients. Then $dX_1, \ldots, dX_r$ are linearly dependent at a point y iff all the r-minors of this matrix vanish at that point. This is clearly a polynomial condition and the set of points at which it is satisfied is a closed set. This proves I.

Hence we may assume that x is semisimple. We may further assume that x belongs to T.

<u>Step II.</u> For a class function $X \in k[G]$, $(dX)_x = 0$ iff $dX_{x/\underline{t}} = 0$, where $\underline{t}$ is the tangent space to T at x. Consider the big cell $U^-.T.U$. Clearly, the tangent space to it at x is the same as the tangent space to G at x. Hence we may consider $U^-.T.U$ instead of G. Let $K = (G_m(k))^r \times \mathbb{A}^s$, s = number of roots. Consider the map $\phi : K \longrightarrow U^-.T.U$ given by: $\phi((t_i), (u_\alpha), (v_\beta))$ $= \prod_{\alpha<0} x_\alpha(u_\alpha) . \prod_i t_i . \prod_{\beta>0} x_\beta(v_\beta)$. This is an isomorphism of varieties by theorem of 3.2. We may consider X to be a function on K (via ϕ). Since ϕ is an isomorphism, we may prove II for K. Any tangent Y to K (at a point y) can be uniquely written as $Y_1 + Y_2$, where Y_1 is the tangent to $(G_m(k))^r$ and Y_2 is a tangent to $\mathbb{A}^s$.

<u>Claim.</u> $(dX)_x(Y) = (dX)_x(Y_1)$. (We identify $\phi(y)$ with y.)

<u>For</u>: X is a polynomial of the form $P_1 + P_2$, where P_1 is a polynomial, each term of which contains <u>at least one</u> of $\{u_\alpha\}_{\alpha<0} \cup \{v_\beta\}_{\beta>0}$ and P_2 is a polynomial in t_i's alone. Since X is a class function, $X(tyt^{-1}) = X(y)$ for each $t \in T$.

If $y = ((t_1), (u_\alpha), (v_\beta))$, then $tyt^{-1} = ((t_1), (\alpha(t).u_\alpha), (\beta(t).v_\beta))$. It is now easy to see that every term of P_1 must contain at least two of $\{u_\alpha\}_{\alpha<0} \cup \{v_\beta\}_{\beta>0}$. Thus, every term of $Y.P_1$ contains at least one of $\{u_\alpha\} \cup \{v_\beta\}$. Since x is given by $((t_1), (0), (0))$, it follows that $(Y.P_1)_x = 0$. Further, it is clear that $(Y_2.P_2)_x = 0$, since P_2 is a polynomial in t_1's alone. Hence, $(Y.X)_x = (Y(P_1 + P_2))_x = (Y.P_2)_x = ((Y_1 + Y_2).P_2)_x = (Y_1.P_2)_x$. Also, $(Y_1.X)_x = (Y_1.(P_1 + P_2))_x = (Y_1.P_2)_x$, since $(Y_1.P_1)_x = 0$.
Hence $(dX)_x\,(Y) = (Y.X)_x = (Y_1.X)_x = (dX_x)\,(Y_1)$.
II follows immediately.

II Shows that it is sufficient to prove that $dX_1, \ldots, dX_r$, restricted to $\underline{t}$, are linearly dependent. Now, $dX_i/_{\underline{t}} = d(X_i/_T)$. In other words, we are concerned with the functions X_i restricted to T only.

On T, $X_i = \lambda_i$ + (terms which are smaller than λ_i) ———— (*)

(See theorem 2 of 3.4).

Lemma. $\prod_{i=1}^{r} dX_i = f.\prod_{i=1}^{r} (\lambda_i^{-1} d\lambda_i)$ (exterior product)

where, $f = \sum_{w\in W} (\det w).(w\delta)$ (= skew δ, by definition)

with $\delta = (\prod_{\alpha>0} \alpha)^{\frac{1}{2}} = \prod_{i=1}^{r} \lambda_i$.

Proof of the lemma. We observe first that $\prod_i \lambda_i^{-1} d\lambda_i$ is not zero. For: the λ_i's are algebraically independent generators of $k[T]$, so that there exist vector fields v_j with $v_j(\lambda_i) = \delta_{ji}$, i.e., with $(d\lambda_i)(v_j) = \delta_{ij}$.

(a) f is skew i.e. $w.f = \det w.f$ for $w \in W$.

For: $w.dX_i = d(w.X_i) = dX_i$, since X_i is a class function.

Hence $w.\prod_i dX_i = \prod_i w.dX_i = \prod dX_i$ ———— (A)

Let $w.\lambda_i = \prod_j \lambda_j^{n_{ij}}$ (written multiplicatively.)

Then $w(\lambda_i^{-1}\, d\lambda_i) = w(\lambda_i^{-1})d(w\lambda_i)$

$$= w\lambda_i^{-1} . d(\prod_j \lambda_j^{n_{ij}})$$

$$= (\prod_j \lambda_j^{-n_{ij}}). (\sum_j (\sum_{k\neq j} \lambda_k^{n_{ik}}). n_{ij}\lambda_j^{n_{ij}-1} . d\lambda_j)$$

$$= \sum_j n_{ij} . \lambda_j^{-1} d\lambda_j .$$

Now, $w.(f. \prod_i \lambda_i^{-1} . d\lambda_i) = wf. \prod_i w(\lambda_i^{-1} d\lambda_i)$

$$= wf. \prod_i (\sum_j n_{ij} . \lambda_j^{-1} . d\lambda_j)$$

$$= (wf) . \det w. \prod_k \lambda_k^{-1} d\lambda_k . \qquad \text{———— (B)}$$

This is so since det w = determinant of (n_{ij}). Thus from (A) and (B), it follows that f = (wf). det w. In other words, wf = det w. f (det w = ± 1).

(b) From *, $dX_i = d\lambda_i + \sum_{\lambda<\lambda_i} d\lambda$

$$= \lambda_i(\lambda_i^{-1}.d\lambda_i) + \sum_{\lambda<\lambda_i} \lambda . (\sum_j m_{ij} \lambda_j^{-1} d\lambda_j), \text{ where } \lambda = \prod_j \lambda_j^{m_{ij}}.$$

It follows that $f = \prod_{i=1}^{r} \lambda_i$ + lower terms. i.e. $f = \delta$ + lower terms.

Since f is skew, an argument similar to one in the proof of theorem 2 of 3.4, gives: $f = \text{skew } \delta + \sum_{\substack{\delta'<\delta \\ \delta' \text{ dominant}}} \text{skew } \delta'$. Consider such a δ'. Since $\delta - \delta'$ = sum of positive roots, it follows that there exists a simple root α_i such that $(\delta - \delta', \alpha_i^*) > 0$. We now have : $1 = (\delta, \alpha_i^*) > (\delta', \alpha_i^*) \geq 0$ and (δ', α_i^*) is an integer. Hence $(\delta', \alpha_i^*) = 0$ i.e. $w_i(\delta') = \delta'$. It now follows that $\det w. w\delta' + \det (w. w_i). w. w_i \delta' = 0$ (det $w_i = -1$). Hence terms in skew- δ' cancel out in pairs, giving skew $\delta' = 0$. Thus f = skew δ, proving the lemma. (This argument holds in case of char k ≠ 2. But the lemma considered may be viewed as a formal identity to be proved in the group algebra of X(T) over $\mathbb{Z}$ from the hypothesis (*). Hence it continues to hold even if char k = 2).

We now use an identity due to Weyl viz. $f = \text{skew } \delta = \delta . \prod_{\alpha > 0}(1 - \alpha^{-1})$. Since x is semisimple and irregular, $\alpha(x) = 1$ for some $\alpha > 0$ (proposition 3 of 3.5). Hence $f(x) = 0$ so that $\prod_i dX_i = 0$. It now follows immediately that $(dX_1)_x, \ldots, (dX_i)_x, \ldots, (dX_r)_x$ are linearly dependent. This proves the implication (2) $\Longrightarrow$ (1).

(1) $\Longrightarrow$ (3). Let $x \in G$ be regular. Pick $y \in C$ such that $X_i(y) = X_i(x) \; \forall \; i$. (This is possible by theorem 5.) As already proved, any element of C is regular ((3) $\Longrightarrow$ (2) $\Longrightarrow$ (1)). Thus, x and y are both regular and belong to the same fibre of p. Hence by theorem 1 above, x and y are conjugate. This proves the implication and the development in theorems 1 to 5.

Problem (Open) Can we find a normal form for non-regular elements in arbitrary simple groups corresponding to the one consisting of several Jordan blocks in SL_n ?

Theorem 7. Let F be a fibre of the map $p: G \longrightarrow \mathbb{A}^r$. (See theorem 1 above.)

(a) The regular elements of F are just the simple ones. (An element in F is simple or non-singular if the dimension of the tangent space to F at that point equals the dimension of F. Such elements form a dense open set in F.)

(b) F is non-singular in codimension 1.

(c) The ideal of F in $k[G]$ (i.e. the ideal of functions in $k[G]$ which vanish on F) is generated by $\{X_i - C_i\}_{1 \leq i \leq r}$ if $F = p^{-1}(C_1, \ldots, C_r)$. Thus the latter ideal is always prime and F is a complete intersection.

(d) F is normal.

Proof. For any variety, the set of simple points is dense in it. In particular, this holds for F. But by theorem 3 of 3.6, the set of regular elements is open and dense in F. Hence there exists a regular element which is simple. Since any two regular elements in F are conjugate, it follows that all the regular elements in F are simple. We shall prove later that the converse of the above statement is also true. This proves (a).

(b) follows immediately from theorem 3 (b) of 3.6.

(c) and (d) : Choose a regular element $x \in F$. By theorem 3, $(dX_1)_x, \ldots, (dX_r)_x$ are linearly independent. Hence $(X_1 - C_1), \ldots, (X_r - C_r)$ form a part of a local co-ordinate system at x. Also, F has codimension r. Now, a general result in algebraic geometry shows that the ideal of F is, in fact, generated by $\{(X_1 - C_1), \ldots, (X_r - C_r)\}$. Since F is irreducible, the ideal of F is prime. From this it can be seen that F is a complete intersection and normal. Since $\{(X_1 - C_1), \ldots, (X_r - C_r)\}$ generate the ideal of F, it follows that the tangent space to F at a point $x (\in F)$ is given by : $\{Y \in (TG)_x / (dX_i)_x (Y) = 0 \ \forall i\}$. Hence $\dim (TF)_x$ = Codimension of the space L generated by $\{(dX_i)_x\}_{1 \leq i \leq r}$. If x is a simple point, then $\dim (TF)_x = \dim F = \dim G - r$. Hence the space L has dimension r. This proves that $\{(dX_i)_x\}_{1 \leq i \leq r}$ are linearly independent. Hence by theorem 3, x is regular. This proves the theorem completely.

Remark. The above applies, in particular, to the variety of unipotent elements.

3.9. Variety of Borel Subgroups. In view of Theorem 7 -(a) of 3.8, our attention is focussed on the irregular unipotent elements i.e. on the singularities of the variety V. The general problem, to which we now turn, is to study these singularities, especially the one at 1.

As a simple example, consider the group $G = SL_2$. An unipotent element is of the form $\begin{bmatrix} a & b \\ c & d \end{bmatrix}$ with $a + d = 2$ and, of course, $ad-bc = 1$. In other words, $V = \left\{ \begin{bmatrix} a & b \\ c & 2-a \end{bmatrix} \middle/ (a-1)^2 = -bc \right\}$. This is clearly a cone in k^3 with vertex at $(1, 0, 0)$. It is now clear that this is the only non-simple point of V. Thus V has exactly one singular point, the vertex of a quadratic cone.

It is pleasant to be able to start with a nice desingularization of V.

Let $\mathcal{B}$ denote the set of all Borel subgroups of a group G. We introduce a variety-structure on $\mathcal{B}$ in the following way: Let G act transitively on a variety V such that (1) the stabilizers are just the Borel subgroups of G. (It is enough to assume that one stabilizer is a Borel subgroup) and (2) for any $v \in V$, the orbit map $G \longrightarrow G.v = V$ is separable (i.e. the differential map is surjective). It now follows that the elements of $\mathcal{B}$ and V are in one-one correspondence ($G_v = G_{v'} \Rightarrow v = v'$, since a Borel subgroup is its own normalizer). We introduce a structure of a variety on $\mathcal{B}$ via this correspondence. We note that $V = G/_B$, B is a fixed Borel subgroup, satisfies the above conditions. It is easy to see that any V with above conditions is isomorphic to one such variety (and hence to any such variety). Thus, we may write $\mathcal{B} = \left\{ gB \in G/_B \right\}$.

Theorem 1. Let V be the variety of unipotent elements in G. Let W be the (closed) subset of $\mathcal{B} \times V$, defined by: $W = \left\{ (B', x)/x \in B' \right\}$. (If B is a fixed Borel subgroup and we have identified $\mathcal{B}$ with G/B, then $W = \left\{ (gB, x)/g^{-1}xg \in U \right\}$.) Then W is a desingularization for V. (We shall define this term presently.)

Proof. A desingularization of a variety V_1 is a pair (V_2, ϕ) where V_2 is a variety and $\phi : V_2 \longrightarrow V_1$ is a morphism such that (1) each of the points of V_2

is simple and (2) $\phi : \phi^{-1}(V_1^s) \longrightarrow V_1^s$ is an isomorphism (V_1^s is the set of simple points in V_1).

Consider the projection $p_2 : W \longrightarrow V$. Our claim is that (W, p_2) is a desingularization of V.

(1) W is irreducible. (It is the image of $G \times U$ under the morphism $\theta: G \times U \longrightarrow G/_B \times V$; $\theta(g,u) = (gB, gug^{-1})$.)

(2) W is non-singular (i.e. each of its points is simple). Consider the big cell $U^-.B$ which is open in G. Let $W' = \{(gB, x) \in W / g \in U^-\}$. Then W' is open in W. Consider the map $\eta : U^- \times U \longrightarrow W'$ given by $\eta(u^-, u) = (u^-B, u^-.u.(u^-)^{-1})$. It is easy to see that η is a morphism and in fact an isomorphism of varieties. (The morphism in the reverse direction is given by: $(g^-B, x) \rightsquigarrow (g^-, (g^-)^{-1}.xg^-)$). Since U^- and U are groups, they are non-singular (A group is homogenous). The product of non-singular varieties is non-singular. It now follows that W' is non-singular. Since W' is open, a point which is non-singular in W' remains so in W as well. Also, translates of W' cover W. It follows immediately that W itself is non-singular.

(3) As seen in theorem 7 of 3.8, the simple points of V are just the regular ones. Let V^r be the set of these points. Let $W^r = p_2^{-1}(V^r)$. We must prove: $p_2 : W^r \longrightarrow V^r$ is an isomorphism. We give the proof only in case of good characteristics. Fix an element $x \in V^r \cap B$. Then the proof of Richardson's theorem shows that the map $\pi : G \longrightarrow V^r$ given by $\pi(g) = gxg^{-1}$ is a quotient map. (The differential is surjective, since $T(Cg^{-1}) = (1-Ad(g))\mathfrak{g}$.) Hence $G/Z_G(x)$ is isomorphic to V^r via π. Consider the morphism $\eta : G \longrightarrow W^r$ given by : $\eta(g) = (gB, gxg^{-1})$. If $g = g'g_0$ with $g_0 \in Z_G(x)$, then $gB = g'.g_0.B = g'B$. (Since x is regular, it belongs to a unique Borel subgroup and here we have $x = g_0.x.g_0^{-1} \in g_0Bg_0^{-1}$.) Hence the morphism η is constant on cosets of $Z_G(x)$ and gives rise to a morphism $\theta : G/Z_G(x) \longrightarrow W^r$. Since $G/Z_G(x)$

is isomorphic to V^r, we have an inverse $V^r \longrightarrow W^r$ of p_2, which is thus an isomorphism. This proves the theorem.

Remarks. (1) For $G = SL_2$, the desingularization of V (the variety of unipotent elements) is isomorphic to $\mathbb{A}^1 \times \mathbb{P}^1$. The picture is  the singular point being blown up to a projective line (G/B).

(2) A generalization of the above theorem is given by the following theorem of Grothendick: The following diagram is a resolution of the singularities of all the fibres of the map $p: G \longrightarrow \mathbb{A}^r$.

$$\begin{array}{ccc} X & \xrightarrow{q} & G \\ {\scriptstyle p'}\downarrow & & \downarrow{\scriptstyle p} \\ Y & \xrightarrow[q']{} & \mathbb{A}^r \end{array}$$

where, $X = \{(B_1, x) \in \mathcal{B} \times G \mid B_1 \ni x\}$

$Y = \{(B, t) \mid B \supseteq T \ni t\}$ $(B, T$ fixed)

and the maps q, p', q' are obtained as follows: $q((B_1, x)) = x; q'((B, t)) = p(t)$. Let $(B_1, x) \in X$ then $B_1 = gBg^{-1}$ for some $g \in G$. Since $x_s \in B_1$, it follows that $g^{-1}x_s g \in B$. Hence $\exists\, b \in B$ such that $b^{-1}g^{-1}.x_s gb \in T$. Also, $gb.Bb^{-1}g^{-1} = B_1$. In other words, $\exists\, h \in G$ such that $B_1 = hBh^{-1}$ and $h^{-1}x_s h = t \in T$. Define $p'((B_1, x)) = (B, t)$. It is easy to check that p' is well-defined and is a morphism. Also, $p(x) = p(x_s) = p(h^{-1}x_s h)$ which gives the commutativity of the above diagram.

We turn to the study of the fibres of the above desingularization of V. Over a fixed u in V we have the variety $\mathcal{B}_u$ of all Borel subgroups containing u, or, on replacing $\mathcal{B}$ by G/B, the variety $(G/B)_u = \{gB \in G/B \mid u \cdot gB = gB\}$ of fixed points of u acting on the "flag variety", hence a projective variety.

Proposition 1. $(G/B)_u$ is connected. (u is an unipotent element.)

<u>Proof</u>. Recall that for a simple root α, P_α denotes the subgroup generated by B and $X_{-\alpha}$. It can be seen that $P_\alpha = (T.\langle X_\alpha, X_{-\alpha}\rangle).U_\alpha$. P_α/B is a projective line. For: If $G_\alpha = \langle X_\alpha, X_{-\alpha}\rangle$ and $B_\alpha = B \cap G_\alpha$, then $P_\alpha/B \backsim G_\alpha/B \sim SL_2/\text{Superdiagonal} \backsim$ projective line. A set $\{xP_\alpha/B,\ x \in G \text{ fixed}\} \subseteq G/B$ is called a line of type α. It is a connected set. It is easy to see that two lines of the same type are either identical or disjoint (since cosets are so). We shall make use of these facts while proving the proposition.

Let $u \in B \cap B_1$. We show that B and B_1 can be connected in $\mathcal{B}_u$ by a sequence of arcs of (projective) lines of various types (*). This clearly proves the proposition.

Let $B_1 = {}^gB$, $g \in G$. By Bruhat lemma, $g \in Bn_wB$. Let $w = w_1 \ldots w_k$, where w_i is the reflection with respect to the simple root α_i and k is minimal with respect to this property. (i.e. $w = w_1 \ldots w_k$ is a 'reduced' expression for w.) We prove (*) by induction on k. If $k = 0$, then $g \in B$ and there is nothing to prove. Let $k \geq 1$. Without loss of generality, we may assume $g = b.n_{w_1} \ldots n_{w_k}$, $b \in B$. Let $g_1 = bn_{w_1} \ldots n_{w_{k-1}}$. Consider the line $g_1.P_{\alpha_k}/B$ of type α_k. This line contains ${}^gB$ and ${}^{g_1}B$. Our claim is that this line is entirely contained in $\mathcal{B}_u$. Granting the claim, we see that $u \in B \cap {}^{g_1}B$. Hence by induction hypothesis, B and ${}^{g_1}B$ can be joined by lines in $\mathcal{B}_u$. Since ${}^{g_1}B$ and ${}^gB$ are joined by a line in $\mathcal{B}_u$ of type α_k (viz. the line $g_1P_{\alpha_k}/B$), the result follows immediately.

We now prove the claim:

(1) By a standard property of reduced expressions, $w_1 \ldots w_{k-1}(\alpha_k) > 0$.

(2) We have, $u \in {}^gB$. Hence $\exists\, u' \in U$ such that $u = {}^gu'$.

Write $u' = \prod_{\substack{\alpha>0 \\ w(\alpha)<0}} x_\alpha(c_\alpha) . \prod_{\substack{\beta>0 \\ w(\beta)>0}} x_\beta(d_\beta)$.

Hence $^{n_w}u' = \prod_{\substack{\alpha>0 \\ w(\alpha)<0}} x_{w(\alpha)}(c'_{w(\alpha)}) \cdot \prod_{\substack{\beta>0 \\ w(\beta)>0}} x_{w(\beta)}(d'_{w(\beta)})$ such that

$c'_{w(\alpha)} = 0$ iff $c_\alpha = 0$ (and also $d'_{w(\beta)} = 0$ iff $d_\beta = 0$). (Here $n_w = n_{w_1} \ldots n_{w_k}$).

Also, $^{n_w}u' = {}^{b^{-1}}u \in U$. Hence $c'_{w(\alpha)} = 0 \; \forall \; w(\alpha) < 0, \alpha > 0$. Since

$w(\alpha_k) = w_1 \ldots w_{k-1} \cdot w_k(\alpha_k) < 0$, it follows that $c_{\alpha_k} = 0$. Thus $u' \in U_{\alpha_k}$. Since

P_{α_k} normalizes U_{α_k}, $^{P_{\alpha_k}}u' \in U_{\alpha_k}$. Hence $^{P_{\alpha_k} g_1^{-1}}u = {}^{P_{\alpha_k} g^{-1}}u =$

$^{P_{\alpha_k}}.u' \in U_{\alpha_k} \subseteq B$. Hence $u \in {}^{g_1 \cdot P_{\alpha_k}}.B$. This proves the proposition completely.

General problem. Study the other properties of the fibres, e.g. their dimensions, number of components etc.

We now prove a proposition which links the dimensions of $Z_G(x)$ and $\mathcal{B}_x$. More precisely, we have:

Proposition 2. Let $x \in G$ be unipotent. Then $\dim Z_G(x) \geqslant r + 2 \cdot \dim \mathcal{B}_x$.

Proof. Let C be the class of x. Let $S = \left\{ (B_1, B_2, y) \in \mathcal{B} \times \mathcal{B} \times C \,/\, y \in B_1 \cap B_2 \right\}$.
For $w \in W$, define $S_w = \left\{ ({}^gB, {}^{gn_w}B, y) \in S \right\}$.
We claim: $S = \bigcup_{w \in W} S_w$, a disjoint union. Consider $(B_1, B_2, y) \in S$. Then there exist $g_1, g_2 \in G$ such that $B_i = {}^{g_i}B$, $i = 1, 2$. By Bruhat lemma, $g_1^{-1} g_2 = bn_w.b'$, $w \in W$. Let $g = g_1.b$. Then ${}^gB = {}^{g_1}B$, ${}^{gn_w}B = {}^{g_2}B$. It now follows that $S = \bigcup_{w \in W} S_w$. Next, let $({}^gB, {}^{gn_w}B, y) = ({}^{g'}B, {}^{g'n_{w'}}B, y)$. Then $g^{-1}g' \in B$ and $n_w^{-1}.g^{-1}.g'.n_{w'} \in B$. Hence $n_{w'} = b.n_w.b'$ for some $b, b' \in B$. Now by Bruhat lemma, $\exists \; z, z' \in Z(T)$ such that $n_{w'} = z.n_w.z'$. In our case (G reductive), $Z(T) = T$ so that $w = w'$. This proves that $S_w \cap S_{w'} = \emptyset$ for $w \neq w'$. It now follows that $\dim S \geqslant \dim S_w$ for all $w \in W$ and the equality holds for at least

one w —— (I).

Considering the projection of $\mathcal{B}\times\mathcal{B}\times C$ onto C and arguing in the same way as lemma 1 of 2.13, we get: $\dim S = \dim C + 2.\dim \mathcal{B}_x$. ——— (II)

(The image is C, while the fibre above x is $\mathcal{B}_x \times \mathcal{B}_x$.)

Fix $w \in W$. S_w can be regarded as a subset of $G/(B \cap {}^{n_w}B) \times C$ via the identification $({}^gB, {}^{gn_w}B, y) \sim ({}^g(B \cap {}^{n_w}B), y)$. Projecting onto the first factor, $\dim S_w = \dim (G/B \cap {}^{n_w}B) + \dim (C \cap U \cap {}^{n_w}U)$. (The fibre above $B \cap {}^{n_w}B$ is $C \cap U \cap {}^{n_w}U$.)

Hence $\dim S_w = \dim G - \dim (B \cap {}^{n_w}B) + \dim (C \cap U \cap {}^{n_w}U)$

$$\leq \dim G - \dim (B \cap {}^{n_w}B) + \dim (U \cap {}^{n_w}U) = \dim G - r.$$

The equality holds iff $C \cap U \cap {}^{n_w}U$ is dense in $U \cap {}^{n_w}U$. ——— (III)

From I, II and III, $\dim C + 2.\dim \mathcal{B}_x = \dim S = \dim S_w$ for some $w \leq \dim G - r$.

Hence $\dim Z_G(x) = \dim G - \dim C \geq r + 2.\dim \mathcal{B}_x$.

This proves the proposition.

Corollary. The equality holds in proposition 2 iff C, the class of x, is dense in some $U \cap {}^{n_w}U$. (i.e. $C \cap U \cap {}^{n_w}U$ is dense in some $U \cap {}^{n_w}U$.)

The proof is clear from (III) in the proposition.

Remark. For each w, there always exists some C with $C \cap U \cap {}^{n_w}U$ dense in $U \cap {}^{n_w}U$ and for elements in this class C, the equality of dimensions above holds.

This gives rise to a problem: Whether the above equality always holds? (for any x and G). The conjecture is that it happens i.e.

Conjecture 1. $\dim Z_G(x) = r + 2.\dim \mathcal{B}_x \ \forall\, x \in G.$

We note that the above fact is easily verified for regular elements, the identity element. It is also known to be true in case of classical groups. It would be enough to prove it for unipotent elements.

Conjecture 2. For a unipotent class C, there exists a $w \in W$ such that $C \cap U \cap {}^{n_w}U$ is dense in $U \cap {}^{n_w}U$.

As seen in the corollary above, conjectures 1 and 2 are equivalent for unipotent elements. Since $U \cap {}^{n_w}U$ is irreducible and hence can contain at most one dense class, these conjectures would imply that the number of unipotent classes is finite, at most $|W|$.

A related conjecture is:

Conjecture 3. For any class, dim C is even. This follows from conjecture 1, since we have the following:

$\dim C = \dim G - \dim Z_G(x) = \dim G - r - 2\cdot\dim \mathcal{B}_x = \dim U^{-}.U - 2\cdot\dim \mathcal{B}_x$ and this is even.

However, conjecture 3 is known to be separately true for good characteristics.

As an example, we consider $G = GL_n$ or SL_n. Here all the conjectures are true since by writing u (unipotent) in normal form we can easily produce a w as in the corollary. (Exercise : do this.) We may verify Conjecture 3 directly as follows: For $x \in G$, write $V = \sum V_i$ with V_i cyclic for x and V_j a a quotient of V_i for $i > j$. Let $n_i = \dim V_i$. If $y \in Z_G(x)$, then y has the form $\begin{bmatrix} B_{11} & B_{12} \cdots & B_{1k} \\ B_{21} & \cdots\cdots & \\ \cdots & \cdots\cdots & \end{bmatrix}$ with B_{ij} an x-module homomorphism of V_j into V_i. The space of such homomorphisms has dimension $\min(n_i, n_j)$. (If $i > j$, then a generator of V_j may go to any element of V_i while if $i < j$ then we apply this to the dual.) Hence $\dim Z_G(x) = \sum_{i,j} \min(n_i, n_j) = n + 2 . \sum_{i<j} \min(n_i, n_j)$ (one less for SL_n). Thus $\dim C(x) = n^2 - n - 2 . \sum_{i<j} \min(n_i, n_j)$, an even number. If there is a single block, then in SL_n, $\dim Z_G(x) = n-1 = r$ so that x is regular as asserted earlier. (See proposition2 of 3.5.) If there are two blocks of size $n-1$ and 1, then $\dim Z_G(x) = r+2$ and x is in the class of "subregular elements" which we shall study presently.

Henceforth, we assume G to be a simple algebraic group.

Theorem 2 (Richardson). Let P be a parabolic subgroup of G. Let U_P be the unipotent radical of P.

(a) There exists in U_P a dense open subset of elements, each of which is contained in a finite number of conjugates of U_P (in G).

(b) ${}^G U_P = \bigcup_{g \in G} g U_P g^{-1}$ is a closed, irreducible subset of dimension = $\dim G - \dim P/U_P$.

(c) If G has a finite number of unipotent classes (e.g. in case of good characteristics), then ${}^G U_P$ contains a unique dense class C of the same dimension as its own. In this case, $C \cap U_P$ is dense in U_P and forms a single class under P.

We observe that these facts have already been proved for $P = B$. (See theorem 3 of 3.6.)

Proof. We may assume that P contains B, a fixed Borel subgroup. We use here some of the standard facts about such parabolic groups. These facts are: P has the following decomposition: $P = M_P . U_P$, where M_P is a reductive group and the product is semi-direct. The root system of M_P can be identified with a subsystem R_P of R generated by simple roots, M_P is generated by T and those X'_αs such that $\alpha \in R_P$. U_P is generated by those X'_αs such that $\alpha \in R^+ - R^+_P$.

We give the proof in several steps:

(1) Let W_P be the Weyl group associated to R_P (i.e. the group generated by $w_\alpha, \alpha \in R^+_P$). Let W^P be the subset of W defined by : $\{w \in W / w(R^+_P) > 0\}$. Then: (a) If $w \in W^P, w' \in W_P$, then $l(ww') = l(w) + l(w')$. (Here $l(w)$ denotes the minimal length of an expression of w as a product of simple reflections)

(b) $W = W^P . W_P$ with uniqueness of expression. This is a standard lemma and we assume it.

(2) $G = \bigcup_{w \in W^P} U_w . n_w . P$.

For: By the reformulation of Bruhat lemma, $G = \bigcup_{w_o \in W} U_{w_o} . n_{w_o} . B$. Thus, it is sufficient to prove that for $w_o \in W, U_{w_o} . n_{w_o} . B \subseteq U_w . n_w . P$ for a suitable $w \in W^P$. By (1) above, $w_o = w . w'$ with $w \in W^P$, $w' \in W_P$. Since $l(w . w') = l(w) + l(w')$, it can be easily checked that $U_{w_o} = U_{w.w'} = U_w . {}^{n_w}U_{w'}$. Hence $U_{w_o} . n_{w_o} . B = U_w . {}^{n_w} U_{w'} . n_w . n_{w'} . B = U_w . n_w . U_{w'} . n_{w'} . B \subseteq U_w . n_w . P$.

(3) For $w \in W^P$, $\dim (U_P \cap {}^w U_P) \leq \dim U_P - l(w)$.

For: $U_P \cap {}^w U_P$ is generated by those X'_α s such that (1) $\alpha \in R^+ - R^+_P$ and (2) $w(\alpha) \in R^+ - R^+_P$. Let K_w denote this set. Since $w(R^+_P) > 0$, $R_w \subseteq R^+ - R^+_P$, where $R_w = \{\alpha > 0 / w(\alpha) < 0\}$ so that $l(w) = |R_w|$. Hence $(R^+ - R^+_P) - K_w \supseteq R_w$. (We note that $R^+ - R^+_P - K_w$ contains β such that $w(\beta) \in R^+_P$, so that equality may or may not hold above). Hence $\dim (U_P \cap {}^w U_P) = |K_w| \leq |R^+ - R^+_P| - |R_w| = \dim U_P - l(w)$.

(4) Let $w \in W^P$ be fixed. Consider the morphism $f_w : U_w \times (U_P \cap {}^w U_P) \longrightarrow U_P$ given by : $f_w(x, y) = xyx^{-1}$. Then U_P contains a <u>dense open</u> set $U'_{P,w}$ such

that $f_w^{-1}(z)$ is finite $\forall z \in U'_{P,w}$.

Proof. Case (i). f_w is dominant. By lemma 2 of 1.13, U_P contains a dense open subset $U'_{P,w}$ such that $\dim f_w^{-1}(z) = \dim (U_w \times (U_P \cap {}^w U_P)) - \dim U_P$, $\forall z \in U'_{P,w}$. By (3), it follows that $\dim f_w^{-1}(z) = 0$ $\forall z \in U'_{P,w}$. In other words, $f_w^{-1}(z)$ is finite.

Case (ii). f_w is not dominant. In this case, let $U'_{P,w} = U_P - \overline{\text{Image of } f_w}$, which is dense open in U_P. Also, $f_w^{-1}(z)$ is empty $\forall z \in U'_{P,w}$. This proves (4).

We are now in a position to prove the theorem.

(a) Let $U'_P = \bigcap_{w \in W^P} U'_{P,w}$, a finite intersection ($U'_{P,w}$ is as in (4)). Then U'_P is dense open in U_P. Let $z \in U'_P$, $w \in W^P$. By choice of $U'_{P,w}$, $z \in U_P \cap {}^{x.w}U_P$ for finitely many $x \in U_w$. Since $G = \bigcup_{w \in W^P} U_w . w . P$, it follows that z belongs to a finite number of conjugates of U_P (in G.)

(b) Consider $S \subseteq G/P \times G$, defined by: $S = \{(gP, x)/g^{-1}xg \in U_P\}$. S is well-defined since P normalizes U_P. Also, S is a closed irreducible subset of $G/P \times G$. (S is the image of $G \times U_P$ under the morphism $(x,y) \rightsquigarrow (xP, xyx^{-1})$.)

By projecting onto the first factor,

$\dim S = \dim G/P + \dim U_P$ (fibres are conjugates of U_P).

By projecting onto the second factor,

$\dim S = \dim {}^G U_P$, since the fibre above any $z \in U'_P$ is finite.

Hence $\dim {}^G U_P = \dim G/P + \dim U_P = \dim G - \dim P/U_P$.

Also, ${}^G U_P$ is closed since S is closed and G/P is complete. This proves (b).

(c) If G has finitely many unipotent classes, then ${}^G U_P$ is made up of finitely many classes. Hence it contains a class C such that $\dim C = \dim {}^G U_P$. Clearly C is dense open in ${}^G U_P$ and hence is unique.

Consider $x \in C \cap U_P$.

$\dim C_P(x) = \dim P - \dim Z_P(x) \geq \dim P - \dim Z_G(x)$

$= \dim P - \dim G + \dim C$

$= \dim P - \dim G + \dim G - \dim P/U_P = \dim U_P$.

Also, $C_P(x) \subseteq U_P \cap C \subseteq U_P$. Hence $\dim C_P(x) = \dim C \cap U_P \ \forall\, x \in C \cap U_P$. It follows that $C \cap U_P$ is a single class under P. Also, by above, $\dim C_P(x) = \dim C \cap U_P = \dim U_P$. Thus $C \cap U_P$ is dense in U_P. This proves theorem completely.

Corollary 1. If $x \in C$ (C is the class mentioned in (c) above), then $\dim Z_G(x) = r + 2.\dim \mathcal{B}_x$.

Proof. Pick $w' \in W_P$ such that $w'(R_P^+) = R_P^-$ (such w' exists). Then $U \cap {}^{w'}U = U_P$.

For: $U \cap {}^{w'}U$ is generated by those X_α's such that $\alpha > 0$ and $w'(\alpha) > 0$. Since $w' \in W_P$, $w'(R^+ - R_P^+) = R^+ - R_P^+ \subseteq R^+$ and $w'(R_P^+) \subseteq R_P^- = R_P^-$. Hence $R^+ - R_P^+$ is precisely the set described above. Hence $U \cap {}^{w'}U = U_P$.

Also, $C \cap U_P$ is dense in U_P. Hence by corollary to proposition 2, $\dim Z_G(x) = r + 2.\dim \mathcal{B}_x$.

Alternate proof. Since $x \in U_P$ and P normalises U_P, it follows that ${}^pB \in \mathcal{B}_x \ \forall p \in P$. Hence $\dim \mathcal{B}_x \geq \dim P/B$.

Now, $\dim P/B = \dim P - \dim B$. Hence $r+2.\dim \mathcal{B}_x \geq r+2.(\dim P - \dim B) = \dim P - \dim U_P$. (This can easily be checked).

Hence $r + 2 \cdot \dim \mathcal{B}_x \geq \dim Z_G(x)$. But $\dim Z_G(x) \geq r + 2 \cdot \dim \mathcal{B}_x$ is always true. Hence $\dim Z_G(x) = r + 2 \cdot \dim \mathcal{B}_x$.

Remark. This eventually proves $\dim \mathcal{B}_x = \dim P/B$.

Corollary 2. If x, C are as in corollary 1, then $Z_G(x)$ is transitive on the conjugates of U_P containing x. (The latter set is finite.)

Proof. Let $x \in U_P \cap {}^gU_P$. Then ${}^{g^{-1}}x,\ x \in U_P \cap C$, which is a single class under P. Hence $\exists\, p \in P$ such that ${}^{p^{-1}g^{-1}}x = x$. Thus $gp \in Z_G(x)$. Also, ${}^{gp}U_P = {}^gU_P$ as P normalizes U_P. This proves the corollary.

3.10. Subregular Elements.

After these generalities, we shall now consider the (unipotent) elements x for which $\dim Z_G(x) = r+2$ (r = rank G). Such elements are called subregular elements. In what follows, we shall prove the existence of subregular (unipotent) elements, show that they form a dense class in the variety of all irregular unipotent elements and then discuss some of their characterizations. We start with the following important lemmas.

Lemma 1. Let G be simple and B a fixed Borel subgroup of G. Fix a maximal torus T and a simple root α.

(1) For any simple $\delta \neq \alpha$ such that $(\delta, \alpha) \neq 0$ and any $x \in U_\alpha$, $\exists\, yB \in P_\alpha/B$ such that $y^{-1}xy \in U_\alpha \cap U_\beta$ ———— (*).

(2) For any simple $\beta \neq \alpha$, define $V_{\alpha,\beta} = \{x \in U_\alpha \,/\, \exists$ precisely $n'_{\alpha,\beta}$ yB's $\in P_\alpha/B$ satisfying (*)$\}$

where $n'_{\alpha,\beta} = \begin{cases} \frac{-2(\alpha,\beta)}{(\alpha,\alpha)} & \text{if this integer is } 0 \text{ or different from char } G. \\ 1 & \text{otherwise.} \end{cases}$

Then $V_{\alpha,\beta}$ is open and dense in U_α.

(3) If $x \in U_\alpha - V_{\alpha,\beta}$ then one of the following possibilities occurs:

(i) Every $yB \in P_\alpha/B$ satisfies (*).

(ii) $\exists\ yB \in P_\alpha/B$ such that $z^{-1}y^{-1}.x.yz \in U_\alpha \cap U_\beta\ \forall\ zB \in P_\beta/B$.

Proof. We first observe that $\{1, x_\alpha(t).n_\alpha\ (t \in k)\}$ is a complete set of representatives of P_α/B (by Bruhat lemma). Let β simple ($\neq \alpha$) such that $(\beta,\alpha) = 0$. We shall prove (2) and (3) in this case. ((1) is not applicable). Let $x = \prod_{\substack{r>0 \\ r\neq\alpha}} x_r(c_r) \in U_\alpha$.

In this case, $n'_{\alpha,\beta} = 0$. Hence $x \in V_{\alpha,\beta}$ iff $\exists$ no $yB \in P_\alpha/B$ satisfying (*). Since $(\beta,\alpha) = 0$, $w_\alpha(\beta) = \beta$ and hence the only positive roots which have support in $\{\alpha,\beta\}$ are α and β. It follows that G_α and X_β commute elementwise. It is now easy to see that for $y \in P_\alpha$, $y^{-1}xy \in U_\beta$ iff $x \in U_\beta$. Hence from above, $x \in V_{\alpha,\beta}$ iff $x \notin U_\beta$ i.e. iff $c_\beta \neq 0$. This shows that $V_{\alpha,\beta}$ is open and dense in U_α. Further, if $x \notin V_{\alpha,\beta}$, then (i) of (3) holds.

We now consider a simple root $\beta \neq \alpha$ such that $(\beta,\alpha) \neq 0$. Let $U_{<\alpha,\beta>} = \prod_{\substack{r>0 \\ \text{Supp}\, r \not\subset \{\alpha,\beta\}}} X_r$. This is a subgroup and is invariant under the conjugation by elements of P_α and P_β. Since our interest lies in actions of P_α and P_β, we may consider $U_\alpha/U_{<\alpha,\beta>}$ instead of U_α. (Note that U_α is a semidirect product extension of $U_{<\alpha,\beta>}$.) Let $R_{\alpha,\beta}$ be the system of roots having support in $\{\alpha,\beta\}$. We may assume $U_\alpha = \prod_{\substack{r>0 \\ r\in R_{\alpha,\beta} \\ r\neq\alpha}} X_r$. Similar assumptions may be made about U_β and $U_\alpha \cap U_\beta$. Also, $R_{\alpha,\beta}$ is a root system with

$\{\alpha, \beta\}$ as basis.

case (i). $R_{\alpha,\beta}$ is of type A_2.

Here, $n'_{\alpha,\beta} = 1$, $U_\alpha = X_\beta . X_{\beta+\alpha}$.

By commutation formulae,

$n_\alpha^{-1} . x_\alpha(t)^{-1} . x_\beta(a) . x_{\beta+\alpha}(b) . x_\alpha(t) . n_\alpha$

$= x_\beta(b+at) . x_{\beta+\alpha}(m_0 . a)$; m_0 : a constant depending on n_α.

Also, $x = x_\beta(a) . x_{\beta+\alpha}(b) \in U_\beta$ iff $a = 0$. Hence if $a = 0$, $1.B \in P_\alpha/B$ satisfies (*) whereas in case $a \neq 0$, $y = x_\alpha(-\frac{b}{a}).n_\alpha$ satisfies (*). Hence (1) follows. To prove (2), we claim: $x \in V_{\alpha,\beta}$ iff $a \neq 0$ or $b \neq 0$. (This clearly defines an open non-empty and hence dense subset of U_α.) Let x be such that $a \neq 0$ or $b \neq 0$. If $a \neq 0$, then $yB = x_\alpha(-\frac{b}{a}).n_\alpha.B$ is the unique solution of (*). If $a = 0$ then $b \neq 0$ and $yB = 1.B$ is the unique solution of (*). Hence $x \in V_{\alpha,\beta}$. Conversely if $a = b = 0$, then every $yB \in P_\alpha/B$ satisfies (*) so that $x \notin V_{\alpha,\beta}$. This proves the claim and also the statement (i) of (3).

case (ii). $R_{\alpha,\beta}$ is of type B_2 and α is the longer toot.

Here, $n'_{\alpha,\beta} = 1$, $U_\alpha = X_\beta . X_{\beta+\alpha} . X_{2\beta+\alpha}$.

In this case, P_α keeps $X_{2\beta+\alpha}$ invariant. Hence this case reduces to case (i) and is proved in a similar fashion.

case (iii). $R_{\alpha,\beta}$ is of type B_2 and α is the shorter root.

Here, $n'_{\alpha,\beta} = 2$ if char $G \neq 2$ and $= 1$ otherwise. $U_\alpha = X_\beta . X_{\beta+\alpha} . X_{\beta+2\alpha}$.

Before proceeding to the proof, we take a closer look at the commutation formulae.

Let $x_\alpha(t)^{-1} . x_\beta(a) . x_\alpha(t) = x_\beta(a) . x_{\beta+\alpha}(mat) . x_{\beta+2\alpha}(m'a\,t^2)$

and $x_\alpha(t)^{-1} . x_{\beta+\alpha}(b) . x_\alpha(t) = x_{\beta+\alpha}(b) . x_{\beta+2\alpha}(m''b\,t)$.

Then it can be proved by iteration that $mm'' = 2m'$ and $m \neq 0, m'' \neq 0$. (A detailed account of such relations may be found in the author's 'Lectures on Chevalley groups, Yale University'.) By changing the parametrizations of $x_{\beta+\alpha}$ $x_{\beta+2\alpha}$, we may arrange $m = 1$, $m' = 1$, so that $m'' = 2$.
Hence choosing n_α properly,

$$n_\alpha^{-1}.x_\alpha(t)^{-1}.x_\beta(a).x_{\beta+\alpha}(b).x_{\beta+2\alpha}(c).\, x_\alpha(t).\, n_\alpha$$

$$= x_\beta(at^2 + 2bt + c).x_{\beta+\alpha}(at + b).x_{\beta+2\alpha}(a)$$

$$= x', \text{ say} \qquad \text{———} \quad \text{I.}$$

Since $w_\beta(\alpha) = \alpha + \beta$ it follows that

$$n_\beta^{-1}.x_\beta(s)^{-1}.\, x'.x_\beta(s).\, n_\beta = x_{-\beta}(at^2 + 2bt + c).x_\alpha(at + b).x_{\beta+2\alpha}(a) \quad \text{———} \quad \text{(II)}.$$

We now turn to the proof of case (iii).

Let $x = x_\beta(a).x_{\beta+\alpha}(b).x_{\beta+2\alpha}(c) \in U_\alpha$.

If $a = 0$, then $1.B$ satisfies (*). If $a \neq 0$, then $x_\alpha(t_0).n_\alpha.B$ satisfies (*), where t_0 is a root of $at^2 + 2bt + c = 0$ (which exists.) Hence (1) is satisfied.

<u>Next, we claim:</u> $x \in V_{\alpha,\beta}$ iff $\Delta = b^2 - ac \neq 0$. (This clearly proves (2).)

We first consider the case when char $G \neq 2$.

Let $a = 0$. Now $\Delta \neq 0$ iff $b \neq 0$. Hence we have to show: $x \in V_{\alpha,\beta}$ iff $b \neq 0$.
It is easy enough to see : $1.B$ and $x_\alpha(\frac{-c}{2b}).n_\alpha.B$ are the two solutions of (*).
Also, $x \notin V_{\alpha,\beta} \Rightarrow b = 0$ and (*) has infinitely many solutions or only one solution according as $c = 0$ or $c \neq 0$. In the first case, (i) of (3) holds. In the second case, $1.B$ is the unique solution and $y^{-1}.x.y = x$. Since $x \in X_{\beta+2\alpha}$ and $(\beta + 2\alpha) \pm \beta$ are not roots, $n_\beta^{-1} x_\beta(s)^{-1}.x.x_\beta(s).n_\beta = x \in U_\alpha \cap U_\beta$ for all s so that (ii) of (3) holds.

Let $a \neq 0$. Then $at^2 + 2bt + c = 0$ has two <u>distinct</u> solutions iff $\Delta \neq 0$. It now follows that $x \in V_{\alpha,\beta}$ iff $\Delta \neq 0$. Further, $x \notin V_{\alpha,\beta} \Rightarrow \Delta = 0$ and $at^2 + 2bt + c = 0$ has a unique solution $t_0 = -\frac{b}{a}$. Since $at_0 + b = 0$, (II) shows that (ii) of (3) holds in this case.

Consider the case when char G = 2. The discussion is similar to one in case char G $\neq$ 2 except that $n'_{\alpha,\beta} = 1$ and the equation $at^2 + 2bt + c$ $(a \neq 0)$ has just one root. This proves this case completely.

case (iv). $R_{\alpha,\beta}$ is of type G_2 and α is the longer root.
Here, $n'_{\alpha,\beta} = 1$, $U_\alpha = X_\beta . X_{\beta+\alpha} . X_{2\beta+\alpha} . X_{3\beta+\alpha} . X_{3\beta+2\alpha}$.
Also, w_α keeps $X_{2\beta+\alpha}$ invariant, takes $X_{3\beta+\alpha}$ into $X_{3\beta+2\alpha}$ and vice-versa. Hence this case is similar to case (i) and is proved in a similar fashion.

case (v). $R_{\alpha,\beta}$ is of type G_2 and α is the shorter root.
Here, $n'_{\alpha,\beta} = 3$ if char G $\neq$ 3 and 1 otherwise. $U_\alpha = X_\beta . X_{\beta+\alpha} . X_{\beta+2\alpha} .$ - $X_{\beta+3\alpha} . X_{2\beta+3\alpha}$.
This case involves calculations which are of the same nature as case (iii). The method used there is applicable here also. We leave the details to the reader.

This proves the lemma completely.

Corollary. If α is any simple root, then ${}^G U_\alpha = \bigcup_{g \in G} g U_\alpha g^{-1}$ is the set of all irregular unipotent elements. (G simple).

Proof. By theorem 1 of 3.7, $x \in U$ is irregular iff $x \in U_\beta$ for some simple root β. Since G is simple, there exists a chain of simple roots $\alpha = \alpha_0, \alpha_1, \ldots, \alpha_k = \beta$ such that $(\alpha_i, \alpha_{i+1}) \neq 0 \ \forall \ 0 \leq i \leq k-1$. By applying the above lemma repeatedly, x can be conjugated into U_α. Since any unipotent element in G can be conjugated into U, it follows that ${}^{G}U_{\alpha}$ is the set of all irregular unipotent elements in G. This proves the corollary.

Lemma 2. Let G, B be as above, α be a simple root.

Define $W_{\alpha} \subseteq U_{\alpha}$ by the following: $x \in W_{\alpha}$ iff

(i) $x \in V_{\alpha,\beta}$ $\forall \beta$ simple $\neq \alpha$.

(ii) For each $yB \in P_{\alpha}/B$ such that $y^{-1}xy \in U_{\alpha} \cap U_{\beta}$, $y^{-1}xy \in V_{\beta,\alpha}$.

(iii) If $\beta \neq \beta'$ are simple roots adjoining α and $yB, y'B \in P_{\alpha}/B$ such that $y^{-1}xy \in U_{\alpha} \cap U_{\beta}, y'^{-1}x.y' \in U_{\alpha} \cap U_{\beta'}$, then $yB \neq y'B$.

Then W_{α} is an open dense subset of U_{α}.

Proof. We give the proof for the case A_r and leave the details of the other cases to the reader. Using the same notation as in lemma 1, let $x = x_{\beta}(a).x_{\beta+\alpha}(b)\ldots$ where β is an adjoining root of α. Then condition (ii) can be seen to be equivalent to $b \neq 0$, which obviously gives an open subset of U_{α}. If $\beta \neq \beta'$ are two adjoining roots of α, then $x \in U_{\alpha}$ may be written as: $x = x_{\beta}(a).x_{\beta+\alpha}(b).-x_{\beta'}(a').x_{\beta'+\alpha}(b')\ldots$ Then the condition (iii) is equivalent to : ($a \neq 0$ or $a' \neq 0$) and $ab' \neq a'b$. This again gives an open subset of U_{α}. It now follows that W_{α} is an open non-empty set of U_{α} (and hence is dense).

Theorem 1. Let G, B be as above.

(a) The set of irregular unipotent elements in G is a closed, irreducible set of codim r+2 in G.

(b) This set contains a unique dense class C of the same dimension as itself. Further, for any simple root α, $C \cap U_{\alpha}$ is dense in U_{α} and forms a single class under P_{α}.

Proof. Fix any simple root α. We apply Richardson's Theorem (of 3.9 - to be referred to as "the theorem" in this proof) to the parabolic subgroup P_α of G. In this case, $U_P = U_\alpha$. The part (b) of "the theorem" and the corollary to lemma 1 above give (a). (We note that dim $P_\alpha/U_\alpha = r+2$.) Assuming G to have only a finite number of unipotent classes (e.g. in good characteristics), part (c) of "the theorem" gives (b) above. However, one can a priori prove that ${}^G U_\alpha$ contains a class of codim r + 2, by a proof to be given elsewhere. Now in the proof of "the theorem", the finiteness of the number of unipotent classes is used only to prove the existence of such a class. Hence (b) above follows even if G has infinitely many unipotent classes. This proves the theorem.

Remarks. (1) The class C above is the set of all subregular unipotent elements.

(2) The two corollaries to Richardson's Theorem hold in the set-up of theorem 1 above.

We now give a development which helps us to give a final characterization of subregular elements.

We recall that a (projective) line of type α (α simple) is a set gP_α/B, $g \in G$. (See Proposition 1 of 3.9.)

Proposition 1. (a) Through any point of G/B passes a unique line of type α.

(b) If u is an unipotent element, then the following statements are equivalent:

(1) u fixes P_α/B pointwise.

(2) $u \in U_\alpha$.

(c) If α, β are distinct simple roots, then the intersection of a line of type β and a line of type α consists of at most one element of $G/_B$. Thus a line of type α is distinct from any line of type β.

(d) If u is a subregular unipotent element, then $Z_G(u)$ acts transitively on the lines of type α, fixed pointwise by u.

Proof. (a) is clear.

(b) (2) $\Rightarrow$ (1) is obvious since P_α normalises U_α.

(1) $\Rightarrow$ (2). It is given that $u.pB = pB \ \forall p \in P_\alpha$. In particular, $u.n_\alpha B = n_\alpha B$ i.e. $n_\alpha^{-1}.u.n_\alpha \in B$. This clearly shows $u \in U_\alpha$.

(c) Let α, β be two distinct simple roots. Consider the lines $xP_\alpha/_B$ and $yP_\beta/_B$. Let xp_1B, $xp_2B(p_1, p_2 \in P_\alpha)$ be two points in their intersection. Then there exist $p_1', p_2' \in P_\beta$ such that $xp_iB = yp_i'B$, $i = 1, 2$. Hence $p_2^{-1} p_1 \in P_\alpha \cap P_\beta$. Using Bruhat lemma, it can be proved that $P_\alpha \cap P_\beta = B$. Hence $p_2^{-1} p_1 \in B$ which shows that $xp_1B = xp_2B$. This proves the required result.

(d) This follows from Theorem 1 (b) of 3.10 and Theorem 2, Cor. 2 of 3.9.

This proves the proposition completely.

Let u be an irregular unipotent element in G. Then as seen in Proposition 1 of 3.9, $(G/_B)_u$ is a connected union of lines of various types. The subregular unipotent element u will now be characterized by the structure of $(G/_B)_u$.

Definition. A Dynkin curve is a non-empty union of lines of various types such that a line of type α meets exactly $n'_{\alpha,\beta}$ lines of type β. (α, β are any distinct simple roots and $n'_{\alpha,\beta}$ is as defined in lemma 1).

Remarks. (a) This notion and a number of its basic properties are due to J. Tits (unpublished).

(b) It is easy to see that the above specifications determine, for each α, the total number of lines of type α viz. $\frac{|\alpha|^2}{|\alpha_{min}|^2}$, where α_{min} is a root of minimum length. The reader may wish to draw pictures of the Dynkin curves of type A_r, B_r, C_r ($r = 3$ is typical), D_r ($r = 4$ or 5), E_6, F_4, G_2. If he does so, he will find that, except for the labeling of the lines, the pictures for G_2 and D_4 are the same, as are those for F_4 and E_6, and similarly for the other cases where roots of different lengths occur.

Proposition 2. Let u be a subregular unipotent element. Then $(G/B)_u$ is a Dynkin curve. Thus, Dynkin curves exist.

Remark. This is the direct analogue of our earlier result (see 3.7) that $(G/B)_u$ is a point if u is regular.

Proof of proposition 2. It can be checked that $(G/B)_{gug^{-1}} = g.(G/B)_u$. Hence one is a Dynkin curve iff the other one is.

Fix a simple root α. Then ${}^G U_\alpha$ = Set of irregular unipotent elements (see the corollary to lemma 1 above). Further, subregular elements in U_α form a single class under P_α and this class is dense in U_α. We use the notation of lemma 1 above. Since $V_{\alpha,\beta}$ is open and dense in U_α, it follows that it contains a subregular element. One may further assume that this element is u itself. Thus we have: u is a subregular element $\in U_\alpha$ such that for any other simple root β, there exist exactly $n'_{\alpha,\beta}$-many points $yB \in P_\alpha/B$ with

$y^{-1}uy \in U_\alpha \cap U_\beta$. Since $u \in U_\alpha$, $P_\alpha/B \subseteq (G/B)_u$ by (b) of proposition 1 above. Let gP_β/B in $(G/B)_u$ intersect this line in a point pB, $p \in P_\alpha$. Now there exists $p' \in P_\beta$ such that $pB = gp'B$. Hence $y = gp' \in P_\alpha$ and $(gp')^{-1}.u(gp') = p'^{-1}.g^{-1}.u.g.p' \in U_\alpha \cap U_\beta$ (Note that $gP_\beta/B \subseteq (G/B)_u \Rightarrow g^{-1}ug \in U_\beta$.) This sets up a one-one correspondence between lines of type β in $(G/B)_u$ which intersect P_α/B and $yB \in P_\alpha/B$ such that $y^{-1}uy \in U_\alpha \cap U_\beta$. It now follows that P_α/B meets exactly $n'_{\alpha,\beta}$ lines of type β in $(G/B)_u$. Let gP_α/B be any other line of type $\alpha \in (G/B)_u$. By part (d) of the proposition 1 above, we may assume $g \in Z_G(u)$. A line hP_β/B of type β intersects gP_α/B if $g^{-1}hP_\beta/B$ intersects P_α/B. Since $g \in Z_G(u)$, $hP_\beta/B \in (G/B)_u$ iff $g^{-1}hP_\beta/B \in (G/B)_u$. It now follows that gP_α/B also meets exactly $n'_{\alpha,\beta}$ lines of type β. This proves that for <u>all</u> u subregular unipotent, every line of type α in $(G/B)_u$ meets exactly $n'_{\alpha,\beta}$ lines of type β in $(G/B)_u$ (β any simple root $\neq \alpha$). Since α is arbitrary, our proposition is proved.

We will prove later that the converse of the above proposition is also true. This characterizes the subregular unipotent elements in G.

<u>Proposition 3.</u> Any two Dynkin curves are translates of each other by elements of G.

<u>Proof.</u> The idea of the proof is as follows: We first give a standard Dynkin curve and then show that any other Dynkin curve is a translate of this curve. This clearly proves the proposition. We give the proof in case G is of type A_r. Using it, proofs in other cases may be given. As an illustration, we give a proof in case G is of type D_r or E_r.

<u>case (i)</u>. Let G be of type A_r.

Let $\alpha_1, \ldots, \alpha_r$ be the simple roots such that $(\alpha_i, \alpha_{i+1}) \neq 0 \ \forall\ 1 \leq i \leq r-1$. Consider the following subset of G/B : $S = P_{\alpha_1}/B \cup n_{\alpha_1} \cdot P_{\alpha_2}/B \cup n_{\alpha_1} \cdot n_{\alpha_2} \cdot P_{\alpha_3}/B \cup \ldots \cup n_{\alpha_1} \ldots n_{\alpha_{r-1}} \cdot P_{\alpha_r}/B$. It is easy to check that S is indeed a Dynkin curve. Let S_1 be any other Dynkin curve. Since S_1 is a non-empty union of lines of various types, it follows that S_1 contains exactly one line of type α_1, say gP_{α_1}/B. Translating S_1 by g^{-1} we find that $g^{-1}S_1$ contains the line P_{α_1}/B. Thus we may assume that S_1 itself contains P_{α_1}/B. Now there exists precisely one line of type α_2 which intersects this line at one of its points. By a suitable element of P_{α_1}, the point of intersection may be translated to $n_{\alpha_1}B$ and this element does not change the line P_{α_1}/B. Hence, again, we may assume that the line of type α_2 occuring in S_1 is $n_{\alpha_1} \cdot P_{\alpha_2}/B$. This meets a unique line of type α_3 in a point $n_{\alpha_1} \cdot x_{\alpha_2}(t) . n_{\alpha_2}B$. (We note that this point cannot be $n_{\alpha_1}B$.) Translating by $b = n_{\alpha_1} \cdot x_{\alpha_2}(t)^{-1} . n_{\alpha_1}^{-1}$, this point is moved to $n_{\alpha_1} \cdot n_{\alpha_2}B$. Also, the lines P_{α_1}/B and $n_{\alpha_1} \cdot P_{\alpha_2}/B$ are kept invariant. Hence we may assume that the unique line of type α_3 occuring in S_1 is $n_{\alpha_1} \cdot n_{\alpha_2} \cdot P_{\alpha_3}/B$. Proceeding in this way, we see that S_1 can be translated to the standard Dynkin curve S. This proves the proposition in this case.

case(ii). Let G be of type E_r or D_r.

Let α_0 be the (unique) branch point. Let $\alpha_1, \ldots, \alpha_k, \alpha_1', \ldots, \alpha_{k'}', \alpha_1'', \ldots, \alpha_{k''}''$ be the other simple roots such that $\{\alpha_1, \ldots, \alpha_k\}$ forms a subsystem of type A_k and similarly for $\{\alpha_1', \ldots, \alpha_{k'}'\}$ and $\{\alpha_1'', \ldots, \alpha_{k''}''\}$. Let S be the standard Dynkin curve for $\{\alpha_1, \ldots, \alpha_k\}$ (see case (i)). Let S', S'' be similarly defined curves for $\{\alpha_1', \ldots, \alpha_{k'}'\}$ and $\{\alpha_1'', \ldots, \alpha_{k''}''\}$ respectively. Consider the following subset S_0 of G/B: $S_0 = P_{\alpha_0}/B \cup n_{\alpha_0} \cdot S \cup x_{\alpha_0}(t_1) . n_{\alpha_0} S' \cup x_{\alpha_0}(t_2) . n_{\alpha_0} \cdot S''$, where $0 \neq t_1, t_2$ and $t_1 \neq t_2$. It is now fairly easy to check that S_0 is indeed a Dynkin curve. Let S_1 be any other Dynkin curve. We may assume that the line of type α_0 occuring in S_1 is P_{α_0}/B. Consider the points of intersection of this line with the lines of type $\alpha_1, \alpha_1', \alpha_1''$ (occuring in S_1). Clearly, these are three distinct points and hence can be translated onto any other triplet of three distinct points, by an element of P_{α_0}. (The action of $P_{\alpha_0}/U_{\alpha_0}$ on P_{α_0}/B is just that of the projective group on the projective line.) Thus we may assume that these three points are $n_{\alpha_0}B$, $x_{\alpha_0}(t_1).n_{\alpha_0}B$, $x_{\alpha_0}(t_2).n_{\alpha_0}.B$ —(*). Now by case (i), the part of S_1, consisting of lines of type $\alpha_1, \ldots, \alpha_k$ is of the form $g.S$, $g \in G$. Also, $n_{\alpha_0}B \neq gn_1B$. (A line of type α_2 does not meet a line of type α_0 in S_1.) Again, the line of type α_1 meeting P_{α_0}/B is $n_{\alpha_0}P_{\alpha_1}/B = gP_{\alpha_1}/B$. It can now be proved that $g = n_{\alpha_0}.b$ with $b \in B$. Let $b = b_1.b_2$, where $b_1 = \prod_{i=1}^{k} x_{\alpha_i}(d_i)$ and $b_2 = \prod_{\substack{r>0 \\ r \neq \alpha_i}} x_r(c_r)$. Then

$n_{\alpha_0}.b_1^{-1}.n_{\alpha_0}^{-1} \in B$. Translation by this element keeps P_{α_0}/B, n_0B, $x_{\alpha_0}(t_i)n_{\alpha_0}B$ $i = 1, 2$ invariant. (e.g. $n_{\alpha_0}.b_1^{-1}.n_{\alpha_0}^{-1}.x_{\alpha_0}(t_1).n_{\alpha_0}B = n_{\alpha_0}.n_{\alpha_0}^{-1}.x_{\alpha_0}(t_1).n_{\alpha_0}.b_1^{-1}B$ as $n_{\alpha_0}^{-1}.x_{\alpha_0}(t_1).n_{\alpha_0}$ commutes with b_1^{-1}). Also, $n_{\alpha_0}.b_1^{-1}.n_{\alpha_0}^{-1}.gS = n_{\alpha_0}.b_1^{-1}.b_1.b_2.S = n_{\alpha_0}b_2S$. Since b_2 does not involve any of the x_{α_i}'s, it follows that $b_2S = S$. Thus by a suitable translation, the Dynkin curve corresponding to $\{\alpha_1, \ldots, \alpha_k\}$ is $n_{\alpha_0}.S$ and the condition (*) is unchanged. Similar arguments for the parts corresponding to $\{\alpha_1', \ldots, \alpha_{k'}'\}$ and $\{\alpha_1'', \ldots, \alpha_{k''}''\}$ show that the whole curve may be translated onto S_0. This proves this case.

<u>case (iii)</u>. G is of type D_r.

Let $\alpha_0, \alpha_1, \ldots, \alpha_{r-1}$ be the simple roots such that $n'_{\alpha_0,\alpha_1} = 2$ and $n'_{\alpha_i, \alpha_{i+1}} = 1 \; \forall \, 1 \leq i \leq r-2$. Let S be the standard Dynkin curve corresponding to $\{\alpha_1, \ldots, \alpha_{r-1}\}$. Let $S_0 = P_{\alpha_0}/B \cup n_{\alpha_0}.S \cup x_{\alpha_0}(t_0).n_{\alpha_0}S$, $t_0 \neq 0$. Then S_0 is the standard Dynkin curve in this case.

<u>case (iv)</u>. G is of type C_r.

Let $\alpha_0, \alpha_1, \ldots, \alpha_{r-1}$ be the simple roots such that $n'_{\alpha_i, \alpha_{i+1}} = 1 \; \forall 0 \leq i \leq r-2$. Let S be the standard Dynkin curve corresponding to $\{\alpha_1, \ldots, \alpha_{r-1}\}$. Then $S_0 = x_{\alpha_1}(t_1).n_{\alpha_1} P_{\alpha_0}/B \cup x_{\alpha_1}(t_2).n_{\alpha_1} P_{\alpha_0}/B \cup S$ where $t_1 \neq t_2, 0 \neq t_1, t_2$, is the standard Dynkin curve.

<u>case (v).</u> G is of type G_2.

Let α, β be the simple roots such that $n'_{\alpha,\beta} = 1$ (i.e. α is the longer root). Then $S_0 = P_\beta/B \cup n_\beta P_\alpha/B \cup x_\beta(t_1).n_\beta . P_\alpha/B \cup x_\beta(t_2).n_\beta P_\alpha/B$ with $0 \neq t_1, t_2$ $t_1 \neq t_2$, is the standard Dynkin curve.

<u>case (vi).</u> G is of type F_4.

Let $\alpha_1, \alpha_2, \alpha_3, \alpha_4$ be the simple roots such that

$n'_{\alpha_1,\alpha_2} = 1$, $n'_{\alpha_2,\alpha_3} = 2$, $n'_{\alpha_3,\alpha_4} = 1$. Then

$$S_0 = n_{\alpha_2} . P_{\alpha_1}/B \cup P_{\alpha_2}/B \cup x_{\alpha_2}(t_1) n_{\alpha_2} . P_{\alpha_3}/B \cup x_{\alpha_2}(t_2) . n_{\alpha_2} . P_{\alpha_3}/B \cup -$$

$$x_{\alpha_2}(t_1) . n_{\alpha_2} . n_{\alpha_3} . P_{\alpha_4}/B \cup x_{\alpha_2}(t_2) . n_{\alpha_2} . n_{\alpha_3} P_{\alpha_4}/B ,$$

where $t_1 \neq t_2$, is the standard Dynkin curve.

<u>Exercise:</u> Work out the proofs for the cases (iii) - (vi).

We are now in a position to give a final characterization of subregular elements.

Let G be a simple group. Choose a simple root α as follows:

(1) arbitrary if G is of type A_r.

(2) the short branch point otherwise. (Omit "short" if all roots have the

same length.)

<u>Theorem 2.</u> Let G, B be as above and u be an irregular unipotent element. Then the following statements are equivalent:

(a) u is subregular i.e. $\dim Z_G(u) = r + 2$.

(b) $(G/B)_u$ (i.e. the variety of Borel subgroups containing u) is a Dynkin curve.

(c) $(G/B)_u$ consists of a union of lines, $\frac{|\beta|^2}{|\beta_{min}|^2}$ of type β, $\forall \beta$ simple. (β_{min} is a root of minimum length.)

(c') u belongs to the unipotent radical of $\frac{|\beta|^2}{|\beta_{min}|^2}$ parabolic subgroups of type β, $\forall \beta$ simple.

(d) $(G/B)_u$ is a finite union of lines of various types.

(d') u is contained in the unipotent radical of a finite number of rank 1 parabolic subgroups.

(e) $(G/B)_u$ has dimension 1.

(f) If α is chosen as above and $u \in U_\alpha$, then $u \in W_\alpha$ (i.e. the conditions (i) - (iii) in Lemma 2 hold.)

<u>Proof.</u> (a) $\Rightarrow$ (b) is precisely proved in Proposition 2 above.

(b) $\Rightarrow$ (c) $\Rightarrow$ (d) $\Rightarrow$ (e) are trivial.

(c') and (d') are reformulations of (c) and (d) respectively.

(e) $\Rightarrow$ (f).

Let, if possible, condition (i) of Lemma 2 fail. Let $x \notin V_{\alpha,\beta}$ for some simple $\beta \neq \alpha$. If $(\beta,\alpha) = 0$, then u fixes $P_\alpha \cdot P_{\beta/B}$ which has dimension 2. This contradicts (e). If $(\beta,\alpha) \neq 0$ i.e. β is adjacent to α, then by lemma 1, either $y^{-1}uy \in U_\alpha \cap U_\beta$ $\forall y \in P_\alpha$ or $\exists y \in P_\alpha$ such that $z^{-1}y^{-1}uyz \in U_\alpha \cap U_\beta$ $\forall z \in P_\beta$. In either case, $(G/B)_u$ contains a subset of dim 2, which is again a contradiction to (e). Thus (i) does not fail. It can be seen similarly that (ii) and (iii) also do not fail. This proves the implication.

(f) $\Rightarrow$ (a). The proof is given case by case.

case (i). G is of type D_r or E_r.
Here, α is the branch point. Let β, γ, δ be the adjacent simple roots.
Let $u = x_\beta(a).x_{\beta+\alpha}(b).x_\gamma(a').x_{\gamma+\alpha}(b').x_\delta(a'').x_{\delta+\alpha}(b'').u'$ where u' does not involve any of the above roots.
Now condition (i) is translated as: If $u' = \prod_\theta x_\theta(c_\theta)$ then $c_\theta \neq 0$ for all simple θ (not adjecent to α). Also, (a,b), (a',b'), (a'',b'') are points of the projective line $\mathbb{P}^1$ (i.e. $(a,b) \neq (0,0)$ etc.) on which G acts. Condition (iii) is translated as: The above three points of projective space are distinct.
Since any triplet of distinct points of $\mathbb{P}^1$ can be taken onto any other triplet of distinct points by an element of SL_2, it follows that we may assume $b=0=a'$, $a'' = b'' \neq 0$ and $a \neq 0, b' \neq 0$. Further, by conjugating by a suitable element $t \in T$, we may assume that $a = 1 = b' = a'' = b''$, $c_\theta = 1 \ \forall \theta$ simple, $(\theta, \alpha) = 0$. Thus our element u looks like: $x_\beta(1).x_{\gamma+\alpha}(1).x_\delta(1).x_{\delta+\alpha}(1).u'$ where $u' = \prod_\theta x_\theta(c_\theta)$ with $c_\theta = 1$ for θ simple, $(\theta,\alpha) = 0$ ——— (*).
Now by lemma 2, W_α is dense in U_α. Also, by Theorem 1, the subregular elements are dense in U_α. Hence there exists a subregular element v which belongs to W_α i.e. satisfies conditions (i) - (iii). Now one may assume that v has the same appearance as given in (*). We claim that u is conjugate to v which proves that u itself is subregular.
Let $U' = \prod_{\substack{\theta > 0 \\ \theta \in I}} X_\theta$, where I excludes all the simple roots and $\beta+\alpha, \gamma+\alpha, \delta+\alpha$.
It follows that $[U_\alpha, U_\alpha] \subseteq U'$ (by commutation formulae). This shows that U_α / U' is abelian. Also, since u and v has the appearance as given in (*),

$uv^{-1} \in U'$. Consider $C_{U_\alpha}(v).v^{-1}$ which contains e and is clearly contained in U'. Since v is subregular, we have: $\operatorname{codim}_{U_\alpha} C_{U_\alpha}(v) = \dim Z_{U_\alpha}(v) \leq \dim Z_G(v) = r + 2$.

Also, $\operatorname{codim}_{U_\alpha} U' = r+2$. $C_{U_\alpha}(v)$ is closed, being an orbit of an unipotent group (see proposition of 2.5). It now follows that $C_{U_\alpha}(v).v^{-1} = U'$ (U' is irreducible). This gives: $uv^{-1} = xvx^{-1}v^{-1}$ for some $x \in U_\alpha$. Hence u and v are conjugate. This proves the implication (f) $\Rightarrow$ (a) in this case.

Other cases are similarly dealt with. We omit the details.

Remark. In the case of D_r and E_r just dealt with, the last inequality must be an equality so that $Z_G(u)^o \subseteq U_\alpha$, u subregular. Hence $Z_G(u)^o$ is unipotent. This result is true for all other cases except A_r and B_r. In these cases, $Z_G(u)^o$ picks up a one-dimensional torus along with the unipotent part. One always has $Z_G(u)^o \subseteq P_\alpha$ (u subregular in U_α) and even $Z_G(u) \subseteq P_\alpha$ in case α is short. In the latter case, for example, $Z_G(u)$ normalizes P_α, hence is part of it.

Final remark. The reader wishing to study conjugacy classes further should consult (2) of the bibliography below, especially Part E.

Appendix. The connection with Kleinian singularities. Let F be a finite subgroup of SU_2 (compact form) acting on $\mathbb{C}^2$. We form the quotient variety $\mathbb{C}^2/F$, a surface S with an isolated, "Kleinian", singularity at the origin 0, coordinatized by the algebra of F-invariant polynomials on $\mathbb{C}^2$. In each case there is a generating set of three polynomials subject to a single relation. The possibilities, up to isomorphism, are as follows:

F	S	Name
Cyclic of order r +1	$xy + z^{r+1} = 0$	A_r
Dihedral of order 2r	$x^{r+1} + xy^2 + z^2 = 0$	D_{r+2}
Binary tetrahedral	$x^4 + y^3 + z^2 = 0$	E_6
Binary octahedral	$x^3y + y^3 + z^2 = 0$	E_7
Binary icosahedral	$x^5 + y^3 + z^2 = 0$	E_8

Since SU_2 acts on $\mathbb{C}^3$, the space of symmetric elements of degree 2 on $\mathbb{C}^2$, as SO_3 on $\mathbb{R}^3$, the kernel being $\pm$ id, the first column may be deduced from the corresponding classification of SO_3 on $\mathbb{R}^3$, as given, e.g., in Weyl's book "Symmetry". In each case the basic invariants and the relation may then be found. For example, if F is cyclic, generated by diag $(\epsilon, \epsilon^{-1})$ $(\epsilon^{r+1} = 1)$ and u, v are the coordinates, then $x = u^{r+1}$, $y = v^{r+1}$, $z = uv$ generate all invariants and satisfy $xy = z^{r+1}$. The other cases are somewhat harder. The third column comes about thus. If $p : S' \longrightarrow S$ is a minimal desingularization, then the fibre $p^{-1}(0)$ above the singular point consists of a union of "lines" (i.e. curves) whose intersection pattern is exactly that of the Dynkin curve with the same lable. We illustrate this desingularization for the type A_r. Here S' is the surface in $\mathbb{C}^3 \times (\mathbb{P}^1)^r$ given by $x = u_1 z,\ u_1 = u_2 z, \ldots, u_{r-1} = u_r z, u_r y = z$ (which imply $xy = z^{r+1}$), and p is the projection into the space of the coordinates x, y, z. That S' is nonsingular and irreducible may be easily proved. We may reverse p on the nonsingular part $S - \{0\}$ of S, e.g. on $x \neq 0$ by solving successively for $u_1, u_2, \ldots,$ so that $p : S' - \{p^{-1}(0)\} \longrightarrow S - \{0\}$ is an isomorphism and we have a desingularization. What does $p^{-1}(0)$ look like? If $u_1 \neq 0$, we can solve successively for $u_2, u_3, \ldots$ (all values ∞), so that we have a line, say L_1, joining $(\infty, \infty, \ldots, \infty)$, or briefly ∞^r, to $(0, \infty^{r-1})$. If $u_1 = 0$ and $u_2 \neq 0$, then $u_3, u_4, \ldots$ in succession all have to be ∞, so that we have a second line L_2 from $(0, \infty^{r-1})$ to $(0^2, \infty^{r-2})$. Similarly we pick up a line L_3 from $(0^2, \infty^{r-2})$ to $(0^3, \infty^{r-3})$, and so on until we have for $p^{-1}(0)$ a Dynkin curve of type A_r. Explicit desingularizations for the other types, especially for E_8, are considerably more complicated, and quite ad hoc. What Proposition 2 above (in conjunction with Theorem 1 of 3.9) means is that this need not be so. Each Kleinian singularity and its desingularization is realized naturally in the corresponding algebraic group, via a "ridge" of singularities on the unipotent variety along its subregular subvariety.

There is a final characterization of subregular elements, analogous to Theorem 3 of 3.8 for regular elements, which we shall mention. This depends on the notion of a deformation of a surface S and a singular point 0 on it. This is a morphism (in some category) of pointed spaces $p: (V, v_0) \longrightarrow (T, t_0)$ such that the fibre $(p^{-1}(t_0), v_0)$ above t_0 is isomorphic to $(S, 0)$. The other fibres represent stages of the deformation as t varies in the base space. The notion of universal deformation can then be defined in an obvious way. Brieskorn has proved the following within the category of germs of analytic spaces (see his talk at Nice, 1970). Let G be simple, simply connected/$\mathbb{C}$, hence a Lie group, u a subregular unipotent element of G, V a subvariety of G (of dim $r+2$) through u transvers to $C(u)$, so that locally G is the product of $C(u)$ and V.

Finally, let $p: G \longrightarrow T/W$ (T max.torus, W Weyl group) be defined by $p(x)$ = set of elements of T conjugate to x_s.

<u>Theorem</u>. Let everything be as just stated.

(a) $p : (V,u) \longrightarrow (T/W, 1)$ is a (the) universal deformation for the corresponding Kleinian singularity.

(b) Conversely, u is subregular unipotent if there exists a factorization

$$\begin{array}{ccc} (G,u) & \xrightarrow{\ q\ } & (X, x_o) \\ {\scriptstyle p}\downarrow & \swarrow {\scriptstyle r} & \\ (T/W, 1) & & \end{array}$$

with q regular and r the just mentioned universal deformation.

Thus we see that the universal deformation takes place naturally within the corresponding algebraic group with the base just the set of semisimple classes by Corollary 2 (a) to Theorem 2 of 2.4.

Consider, for example, the group SL_2. Here $u = 1$, V is SL_2 itself, and $p(x)$ is basically just the trace of x. If $x = \begin{bmatrix} a & b \\ c & d \end{bmatrix}$, then $ad - bc = 1$ so that $-bc = (a - \frac{p(x)}{2})^2 + (1 - \frac{p(x)^2}{4})$. At $x = 1$, $p(x) = 2$, we have a cone with the singularity of type A_1 at the vertex. Nearby $p(x) \neq 2$, the surface is an ellipsoid, and the singularity has disappeared.

REFERENCES

(1) Armand Borel : Linear Algebraic Groups. (W.A. Benjamin Inc. 1969.)

(2) ——————— et al.: Seminar on Algebraic Groups and Related Finite Groups. (Springer-Verlag Series No. 131.)

(3) Nicolas Bourbaki : Groupes et Algebres de Lie-Chapitre 4, 5 et 6. (Hermann, Paris 1968.)

(4) Claude Chevalley : Seminaire Chevalley Vol. 1, 2.

(5) David Mumford : Introduction to Algebraic Geometry. (Harvard University Notes.)

(6) Jean-Pierre Serre : Lie Algebras and Lie Groups. (Harvard University Notes, W.A. Benjamin Inc. 1965.)

(7) ——————— : Algebres de Lie Semisimples Complexes (W.A. Benjamin Inc. 1966.)

(8) Robert Steinberg : Regular elements of Semisimple Algebraic Groups. (I.H.E.S. Publi. Mathe. No. 25, 1965.)

(9) ——————— : Lectures on Chevalley Groups. (Yale University Notes, 1967.)

(10) ——————— : Endomorphisms of Linear Algebraic Groups. (A.M.S. Memoirs No. 80, 1968.)